Photosynthesis in the Marine Environment

Photosynthesis in the Marine Environment

Sven Beer, Mats Björk and John Beardall

WILEY Blackwell

This edition first published 2014 © 2014 by John Wiley & Sons, Ltd.

Editorial offices: 2121 State Avenue, Ames, Iowa 50014-8300, USA
                     The Atrium, Southern Gate, Chichester, West Sussex, PO19 8SQ, UK
                     9600 Garsington Road, Oxford, OX4 2DQ, UK

For details of our global editorial offices, for customer services and for information about how to apply for permission to reuse the copyright material in this book please see our website at www.wiley.com/wiley-blackwell.

Library of Congress Cataloging-in-Publication Data

Beer, Sven, 1949–
    Photosynthesis in the marine environment / Sven Beer, Mats Björk, and John Beardall. – Second edition.
        pages cm
    Includes bibliographical references and index.
        ISBN 978-1-119-97958-6 (cloth) – ISBN 978-1-119-97957-9 (pbk.)   1. Photosynthesis.   2. Plants–Effect of underwater light on.   3. Aquatic plants–Ecophysiology.   4. Underwater light.   I. Björk, Mats.   II. Beardall, John.   III. Title.
        QK882.B35 2014
        581.7′6–dc23

                                                                                          2014002673

A catalogue record for this book is available from the British Library.

Front cover - A thick turf of red and brown algae growing in the high-energy surf zone, Malta. Photo by Mats Björk
Back cover (from left to right) - Colonies of the marine cyanobacteria Trichodesmium (see Figure 2.1b). Photo by Birgitta Bergman - Quantum yield measurement of zooxanthellae in a coral by PAM fluorometry (see Figure 9.7). Photo by Katrin Österlund - Various macroalgae in the Mediterranean, with the brown alga Padina sp. on top. Photo by Katrin Österlund - The seagrass Cymodocea serrulata, Zanzibar. Photo by Mats Björk.

Cover design by Steve Thompson

Set in 11/13pt Minion by Aptara Inc., New Delhi, India
Printed and bound in Malaysia by Vivar Printing Sdn Bhd

1   2014

Contents

About the authors

Sven Beer
(svenb@ex.tau.ac.il)

Mats Björk
(mats.bjork@su.se)

John Beardall
(john.beardall@monash.edu)

Prof. Sven Beer was, as his first name indicates, born in Sweden (in 1949), where he also completed his BSc in Biology at Stockholm and Uppsala Universities. After moving to Israel in 1973, he graduated with a PhD (title of dissertation: Photosynthesis of Marine Angiosperms) from Tel Aviv University, where he has worked in teaching and research ever since; he is now a Professor Emeritus at that university, and also holds an affiliated professorship at Stockholm University. During his career, Sven has authored some 120 scientific publications in the field of marine botany and, especially, photosynthesis of marine macrophytes. You can read more about Sven on his home page www.tau.ac.il/lifesci/departments/plant_s/ members/beer/beer.html

Prof. Mats Björk, was born in Stockholm, Sweden in 1960. He moved to Uppsala for university studies, and earned his doctorate degree there in 1992. In 1996 he took up a research position at Stockholm University, where he currently is Professor of Marine Plant Physiology. His research focuses on marine plants and their productivity in both tropical and temperate environments, and how they are affected by environmental change such as pollution and ocean acidification. He has worked extensively with academic institutions in Africa, where he has co-ordinated various educational and research projects and guided many graduate students.

Prof. John Beardall was born in the UK in 1951. He studied Microbiology at Queen Elizabeth College, University of London, prior to moving across town to University College (also University of London) for his PhD. Following post-doctoral studies at University College of North Wales, then at University of Dundee, John moved to Australia in 1982, taking up a Lecturer position at La Trobe's Department of Botany. He and his team moved to Monash in 1988 when John was appointed Senior Lecturer in the, then, Department of Botany. He currently holds a Professorial position in the School of Biological Sciences. John has published 130 papers and over 20 book chapters. His research group focuses on the physiology of algae in relation to environmental factors. A major interest is related to understanding the ways in which marine and freshwater microalgae, including the cyanobacteria responsible for toxic blooms in inland and coastal waters, will be influenced by global change. John's home page is http://monash.edu/science/about/schools/biological-sciences/staff/beardall/

Contributing authors

John Raven (p. 7), University of Dundee, UK (j.a.raven@dundee.ac.uk)
Birgitta Bergman (p. 18), Stockholm University, SE (birgitta.bergman)@su.se
Laura Steindler (p. 25), Haifa University, IL (lsteindler@univ.haifa.ac.il)
Gidon Winters (p. 102), The Dead Sea – Arava Science Center, IL (wintersg@adssc.org)
Alvaro Israel (p. 109), Israel Oceanographic and Limnological Research, IL (alvaro@ocean.org.il)
Lennart Axelsson (p. 112), Kristineberg Marine Research Station, SE (alglax@telia.com)
Stewart Larsen (p. 148), Monash University, AU (stuart.larsen@monash.edu)
Yoni Sharon (p. 172), MBD Energy Ltd., AU (2yonisharon@gmail.com)
Aaron Kaplan (p. 190), The Hebrew University, IL (aaronka@vms.huji.ac.il)
Rui Santos (p. 194), University of Algarve, PT (rosantos@ualg.pt)

Preface

When we started to write this book, there were already a few excellent textbooks on the market describing photosynthesis in aquatic, including marine, environments. Noteworthy among them are the graduate-level books Aquatic Photosynthesis by our friends and colleagues Paul Falkowski and John Raven and the equally high-level book Light and Photosynthesis in Aquatic Ecosystems by our more recent acquaintance John Kirk. In addition, Anthony Larkum, Susan Douglas and John Raven had edited a multi-authored volume in the series Advances in Photosynthesis and Respiration entitled Photosynthesis in Algae. So why, then, yet another book on marine photosynthesis? First, the present text is suitable also for under-graduate level university and college students (BSc and equal levels) of biology or marine sciences, and parts can also be used for high-school science students who wish to brush up on photosynthesis and connect it with marine studies. Secondly, while the previous books describe marine photosynthesis and light as its driving force in great detail, the present book takes a more general look upon our oceans as an important habitat for plant growth (using the broad definition of marine 'plants' as including all photosynthetic organisms in the oceans), and we try in our text to link photosynthesis to the very special and different-from-terrestrial milieus that marine environments provide. Thus, researchers in the fields of terrestrial plant, or more general

marine, sciences may here also be introduced to, or complement their knowledge in, the expanding field of marine plant 'ecophysiology' as related specifically to photosynthesis as the basis for marine plant growth. In this context we treat acclimations not only to light and the inorganic carbon composition of seawater, but also, e.g. to desiccation in the intertidal and pH in rockpools. Also, there is a chapter on how some marine plants can, through their photosynthetic characteristics, change the environment for other plants. Further, the present book fills in some gaps not covered extensively in the other volumes. These include subjects such as photosymbiosis in marine invertebrates, significantly larger portions devoted to marine macrophytes (macroalgae and, especially, seagrasses) and a background to common methodologies used for marine photosynthetic research. Lastly, but most importantly, however, the present book came about because we felt the need to convey our experiences in teaching and working on various aspects of marine plant photosynthesis to a larger audience than those peers and graduate students that (occasionally) read our scientific papers. The fact that much of this book reflects those experiences also goes hand in hand with the fun we had writing it; we hope you will enjoy reading it too!

Sven Beer, Mats Björk and John Beardall
Tel Aviv, Stockholm and Melbourne

PS: This is a First Edition and, so, mistakes are unavoidable. In order to make forthcoming editions better, we would be happy for any advice on improvements. Therefore, please do e-mail us any comments you may have on the contents of this book.

Cover photo: A thick turf of red and brown algae growing in the high-energy surf zone, Malta. Photo by Mats Björk.

The green alga *Caulerpa racemosa* at Zanzibar. Photo by Katrin Österlund.

About the companion website

This book is accompanied by a companion website:

www.wiley.com/go/beer/photosynthesis

The website includes:

- Powerpoints of all figures from the book for downloading
- PDFs of tables from the book

The brown alga *Fucus serratus* at the Swedish west coast. Photo by Mats Björk.

Part I
Plants and the Oceans

Introduction

Our planet as viewed from space is largely blue. This is because it is largely covered by water, mainly in the form of oceans (and we explain why the oceans are blue in Box 3.2, i.e. Box 2 in Chapter 3). So why then call this planet **Earth?** In our view, Planet Ocean, or **Oceanus**, would be a more suitable name; since the primary hit on Google for 'Planet Ocean' is a watch brand, let's meanwhile stick with Oceanus. Not only is the area covered by oceans larger than that covered by earth, rocks, cities and other dry places, but the volume of the oceans that can sustain life is vastly larger than that of their terrestrial counterparts. Thus, if we approximate that the majority of terrestrial life extends from just beneath the soil surface (where bacteria and some worms live) to some tens of metres (m) up (where the birds soar), and given the fact that, on the other hand, life in the oceans extends to their deepest depths of some 11 000 m, then a simple multiplication of the surface areas (~70% for oceans and ~30% for land) with the average depth of the oceans (3800 m), or height of terrestrial environments, gives the result that the oceans constitute a life-sustaining volume that is some 1000 times larger than that of the land. If we do the same calculation for plants only under the assumptions that the average terrestrial-plant height is 1 m (including the roots) and that there is enough light to drive **diel positive apparent** (or **net**) **photosynthesis**[1] down to an average depth of 100 m, then the 'life volume' for plant growth in the oceans is 250 times that provided by the terrestrial environments.

It has been estimated that photosynthesis by **aquatic plants**[2] provides roughly half of the

---

[1] Diel (= within 24 h, i.e. during the day AND the night) apparent (or net) photosynthesis is the **metabolic gas exchange of CO_2 or O_2** resulting from photosynthesis and respiration in a **24-h cycle** of light (where both processes take place) and darkness (where only respiration occurs). Net photosynthesis is positive if, during these 24 h, there is a net consumption of CO_2 or a net production of O_2.

[2] The word '**plant(s)**' is used here in the wider sense of encompassing all photosynthetic organisms, including those algae sometimes classified as belonging to the kingdom Protista (in, e.g. Peter Raven's et al. textbook Biology of Plants) and, unconventionally, also those photosynthetic prokaryotes referred to as cyanobacteria. The latter are of course not included taxonomically in the kingdom Plantae (they belong to the Eubacteria), but can, just like the algae, functionally be considered as 'plants' in terms

Photosynthesis in the Marine Environment, First Edition. Sven Beer, Mats Björk and John Beardall.
© 2014 John Wiley & Sons, Ltd. Published 2014 by John Wiley & Sons, Ltd.
Companion Website: www.wiley.com/go/beer/photosynthesis

global primary production[3] of organic matter (see, e.g. the book Aquatic Photosynthesis by Paul Falkowski and John Raven). Given that the salty seas occupy much larger areas of our planet than the freshwater lakes and rivers, there is no doubt that the vastly greater part of that aquatic productivity stems from photosynthetic organisms of the oceans. Other researchers have indicated that the marine primary (i.e. photosynthetic) productivity may be even higher than the terrestrial one: Woodward in a 2007 paper estimated the global marine primary production to be 65 Gt[4] carbon year^{-1} while the terrestrial one was 60 Gt year^{-1}. This, and recently increasing realisations of the important contributions from very small, cyanobacterial, organisms previously missed by researchers (see Figure I.1a), lends thought to the realistic possibility that photosynthesis in the marine environment contributes to the major part of the primary productivity worldwide. Some of the players contributing to the marine primary production are depicted in Fig. I.1, while many others are described in Chapter 2.

Because marine plants are the basis for generating energy for virtually every marine food web[5], and given that the oceans may be even more productive than all terrestrial environments together, it is logical that the process of marine photosynthesis should be of interest to every biologist. But what about non-biologists? Why should the average person care about marine photosynthesis or marine plants? One of SB's in-laws has never eaten algae or any other products stemming from the sea; he hates even the smell of fish! However, after telling him that the oxygen (O_2) we breathe is (virtually, see Box 1.2) exclusively generated by plants through the process of photosynthesis, and that approximately half of the global photosynthetic activity takes place in the seas, even he agreed that any interference with the oceans that would lower their photosynthetic production would indeed also jeopardise his own wellbeing. And for those who like fish as part of their diet, the primary production of **phytoplankton**[6] is directly related to global fisheries' catches. So, for whatever reason, the need to maintain a healthy marine environment that promotes high rates of photosynthesis and, accordingly, plant growth in the oceans should be of high concern for everyone, just as it is of more intuitive concern to maintain healthy terrestrial environments.

This book will in this, its first, part initially review what we think we know about the evolution of marine photosynthetic organisms.

---

of their (important) role as primary (photosynthetic) producers in the oceans.

[3]Since photosynthesis is the basic process that produces energy for the life and growth of all organisms, its outcome is often called **primary production**; the rate at which the process of primary production occurs is called **primary productivity**.

[4]In the article by Woodward, the amounts of carbon is given in Pg (petagrams, $=10^{15}$ g). However, here we will adhere to the equally large unit Gt (gigatonnes, $=10^9$ tonnes) for global production rates of plants.

[5]The word 'virtually' is inserted as a caution from making too broad generalisations, and there are exceptions (which is true for almost every aspect of biology): It is for instance possible that runoff containing organic matter from land could partly sustain some near-shore ecosystems. Also, non-photosynthetic formation of organic matter (in a process called chemolithotrophy, where, e.g. energy to drive CO_2 assimilation is obtained from the oxidation of hydrogen sulphided to sulphur) near hydrothermal vents in the deep ocean has been found to feed entire, albeit quite space-limited, ecosystems.

[6]**Phytoplankton** (phytos = plant; plankton = 'drifters') refers here to all photosynthetic organisms that are carried around by the waves and currents of the oceans, be it small, often unicellular, cyanobacteria, eukaryotic microalgae or the occasional larger alga drifting in the water body. In the animal kingdom, the analogues to phytoplankton are the zooplankton; those animals that can direct their way by swimming against the currents are called nekton (e.g. most adult fish and other larger forms).

Figure I.1   Some of the 'players' in marine photosynthetic productivity. (a) The tiniest cyanobacterium (*Prochlorococcus marinus*) – the cyanobacteria may provide close to half of the oceanic primary production, at least in nutrient-poor waters. (b) A eukaryotic phytoplankter (the microalga *Chaetoceros* sp.) – the microalgae are probably the main primary producers of the seas. (c) Two photosymbiont- (zooxanthellae-)containing corals from the Red Sea (*Millepora sp.*, left, and *Stylophora pistillata*, right) – while quantitatively minor providers to global primary production, these photosymbionts keep the coral reefs alive. (d) The temperate macroalga *Laminaria digitata* – macroalgae provide maybe up to 10% of the marine primary production. (e) A meadow of the Mediterranean seagrass *Posidonia oceanica* – even though their primary productivity is amongst the highest in the world on an area basis, the contribution of seagrasses to the global primary production is small (but they form beautiful meadows!). Photos with permission from, and thanks to, William K. W. Li (Bedford Institute of Oceanography, Dartmouth, NS, Canada) and Frédéric Partensky (Station Biologique, CNRS, Roscoff, France) (a), Olivia Sackett (b), Sven Beer (c), Katrin Österlund (d) and Mats Björk (e).

Personal Note: Why algae are important

Most of the students taking my (JB) Marine Biology class are more interested in the animals (especially the 'charismatic' mega-fauna like dolphins, turtles and whales) than in those plants that ultimately provide their food. Indeed, when I first started my PhD, my supervisor advised me to "...never admit at parties to what you are studying (i.e. algae), or no-one will talk to you...". However, algae (as a major proportion of the marine flora) are far more important than 'just those things that cause green scum in swimming pools'. I deal with this in the very first lecture of my course by asking the students to put down their pens (or these days their laptops) and take a deep breath. They do this and I then I ask them to take a second deep breath – by this time they are wondering if I've finally lost it, but then I explain that the second breath they have just taken is using oxygen from photosynthesis by marine plants and that if it wasn't for this group of organisms, not only would our fisheries be more depleted than they already are, but also we'd only have half the oxygen to breath! This tends to focus their attention on those organisms at the bottom of the marine food chain and the fact that algae are responsible for around half of the biological CO_2 draw-down that occurs on our planet. As a consequence, algae get more of their respect.

—John Beardall.

Then, the different photosynthetic 'players' in the various marine environments will be introduced: the cyanobacteria and microalgae that constitute the bulk of the phytoplankton in sunlit open waters, the photosymbionts that inhabit many marine invertebrates, and the different macroalgae as well as seagrasses in those benthic[7] environments where there is enough light during the day for dielly positive net (or 'apparent', see, e.g. Section 8.1) photosynthesis to take place. Finally, the third chapter of this part will outline the different properties of seawater that are conducive to plant photosynthesis and growth, and in doing so will especially focus on what we view as the two main differences between the marine and terrestrial environments: the availability of **light** and **inorganic carbon**[8]. In doing so, this first part will, hopefully, become the basis for understanding the other two parts, which deal more specifically with photosynthesis in the marine environment.

---

[7] Benthic refers to the environments close to the bottom of water bodies (as opposed to the water column itself); the flora and fauna of the **benthos** are those organisms living in that environment and they include all sessile (bottom-attached) plants, the **phytobenthos**, as well as sessile and other bottom-dwelling animals such as cod.

[8] Unlike terrestrial plants, which use only atmospheric CO_2 for their photosynthesis, **marine plants can also utilise other inorganic carbon (Ci) forms** dissolved in seawater (see Chapters 3 and 7).

Chapter 1

The evolution of photosynthetic organisms in the oceans

The Big Bang started our universe some 14.7 billion years ago, but our planet, Earth (or Oceanus as we like to call it), was formed only 4.5 billion years ago … making it possible that life started elsewhere in the universe[1] (see the below box by John Raven). In any case, fossil evidence[2] suggests that the first life forms appeared on Earth about 1 billion years after its formation, i.e. ~3.5 billion years ago. If we favour the theory that life started on Earth (rather than on another planet), then it is commonly thought, or at least taught, that the organic compounds that eventually led to the living cells formed at the time when water vapour condensed as our planet cooled. The atmosphere was then anoxic (no molecular oxygen, O_2, was present) but contained high concentrations of carbon dioxide (CO_2) and methane (CH_3), as well as nitrogen gas (N_2), and those same conditions were also present

[1]The fact that no one was around to witness the evolution of living organisms on Earth renders deductions only of the evolution of life as based on fossil findings[2], and are influenced also by a dust of logics and/or imagination. Also, given the enormous amount of planets orbiting the enormous numbers of stars in the (possibly) infinite number of galaxies (and perhaps also universes, i.e. our universe may be part of a multiverse), lends thought to the, we think very realistic, likelihood that life formed also elsewhere in parallel with, or before or after, life on Earth, lending thought to the possibility that primitive life forms were brought to Earth from other planets by meteorites that impacted the early Earth. These primitive organisms then evolved on our planet during the last 3.5 billion years. (See also the following box by John Raven.)

[2]We note that ~3.5-billion-year old fossils of 'bacteria-like' life forms were recently found in sandstones at the base of the Strelley Pool rock formation in Western Australia. However, such fossils often appear as rather fuzzy and must be hard to interpret as representing this or that organism.

Photosynthesis in the Marine Environment, First Edition. Sven Beer, Mats Björk and John Beardall.
© 2014 John Wiley & Sons, Ltd. Published 2014 by John Wiley & Sons, Ltd.
Companion Website: www.wiley.com/go/beer/photosynthesis

in the ponds that were in equilibrium with the atmosphere (see Box 1.1 for a glimpse into air-water equilibria), or the haze (see below), in which the first organic molecules were formed. Under those conditions, it is thought that simple organic compounds (sugars, amino acids and organic acids) were formed by chemically combining water (H_2O) with the dissolved gases CO_2, N_2 and possibly hydrogen (H_2), methane (CH_4), ammonia (NH_3) and hydrogen sulphide (H_2S). (We note with interest that C, H, O and N make up >95% of the elemental content in living organisms.) In the absence of photosynthesis (which probably developed later), the energy for forming these organic compounds came not from sunlight but from lightning, high levels of UV radiation and possibly also radioactivity, or a combination of those energy sources. That lightning could cause such gases to combine into simple organic compounds in an aquatic medium

was shown experimentally in the 1950s: Stanley Miller and Harold Urey (and co-workers) showed then that organic compounds could be formed from simpler compounds believed to be present in Earth's atmosphere some 3.5 billion years ago. This was done by introducing the gases CH_4, NH_3 and H_2 into a loop of water vapour generated by boiling water; when the gas mixture passed through an electrical discharge, organic molecules were formed, and these accumulated in a cooled water trap at the bottom of the apparatus (representing the droplet, pond or ocean where life originated, see Figure 1.1). Subsequent complementary experiments have confirmed that up to 20 different amino acids, and small proteins, can be formed in a similar fashion. Amino acids were, however, also found in a meteorite that fell to Earth in 1969 (see Box 1.2 by John Raven for a discussion on alternative sites to our planet for the origins of life).

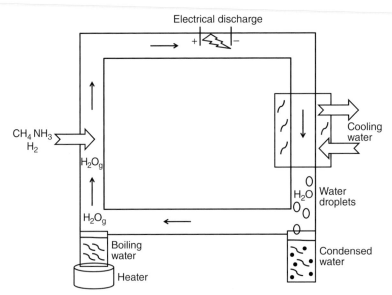

Figure 1.1   The Miller and Urey experiment. Gases assumed to be present in the primordial atmosphere 3.5 billion years ago were introduced into a closed glass apparatus in which they were carried in a loop of water vapour (H_2O_g) generated from boiling water (lower left). After passing through electrical discharges, the gases were cooled, and the droplets formed by condensation filled up a cup (lower right). Organic compounds (black dots) were then identified in the collected condensate. Drawing by Sven Beer.

Box 1.1   Air–water equilibria

It is unfortunate that **atmospheric** gas compositions are usually reported when discussing chemical and biochemical reactions; at least the latter always take place in **liquid media**! While the relative concentrations of dissolved gases correlate directly with those of the atmosphere (that they are in equilibrium with) according to Henry's law (e.g. a doubling of O_2 in the gas phase will cause a doubling of dissolved O_2 at air–water equilibrium at a set temperature), the absolute concentrations of gases in those two media may differ greatly. For example, while the concentration of CO_2 in seawater water is about 75% of that in air at 20 °C (on a mol per volume, or molar, basis[3]), that of O_2 is only 2.6% of that in air. Thus, the O_2 concentration in air is rather irrelevant when discussing, e.g. rates of respiration in marine organisms. Also for respiration in terrestrial organisms it is rather irrelevant to discuss biochemical processes with respect to aerial O_2 concentrations since there, too, they take place in the liquid medium of the cells (or, rather, liquid-filled cellular compartments where they occur). Thus, there may be little difference in concentrations of dissolved gases in a huge-volume oceanic water body and a small-volume cell; the latter may be seen as being a tiny ocean. (Unlike **concentrations** in air-equilibrated water bodies, there **is** a difference in **diffusivity** of, e.g. gases in small and large water bodies, but more about that in Chapter 3).

[3]Molarity will mostly be used in this book when referring to dissolved compounds (including gases). In the gas phase, percentages (or parts per million, ppm) can easily be converted to molarity with the knowledge that 1 mol of any gas equals 22.4 litres (L) (at normal conditions): micromol L^{-1} (or micromolar, μM) thus equals ppm/22.4.

Box 1.2   Life in Far Away Places

by **John Raven**, University of Dundee, UK (j.a.raven@dundee.ac.uk)

Astrobiology has developed over the last few decades as the science that considers the possibility of life elsewhere in the Universe. The subject originated in earlier speculations about life elsewhere, with such concepts as panspermia. The subject has become more firmly grounded since the discovery of planets orbiting stars other than our sun: in mid-July 2013 there are about 900 (depending on the source consulted) well-authenticated reports of exoplanets.

Consideration of life elsewhere in the universe is inevitably based on our understanding of life on Earth, i.e. LAWKI (Life As We Know It), and the limits of its occurrence. LAWKI depends on water, hence NASA's mantra "Follow the Water". The occurrence of liquid water on a planet depends on the radiant energy emitted by the parent star and the orbital distance of the planet, as modulated by factors such as the greenhouse effect of the planetary atmosphere. The solar energy output determines the habitable zone, i.e. the range of orbital distances at which liquid water can occur. Since the energy output of stars increases with time, the habitable zone moves out from the star and this gave rise to the concept of the continuously habitable zone, a narrower range of distances from the star at which liquid water could occur over times (hundreds of millions to billions of years) consistent with the origin and, particularly, the evolution of life.

Another aspect of astrobiology is the concept of the Earth-like planet, i.e. a rocky planet large enough to retain an atmosphere rather than a gas giant. The available methods for detecting exoplanets are all biased toward larger planets with small orbital distances, i.e. well inside the continuously habitable zone, so that most of the known exoplanets are not Earth-like planets in the continuously habitable zone. However, with improved means of detecting exoplanets it is likely that many more exoplanets capable of supporting life will be detected over the coming decades.

The possibility of supporting life does not necessarily mean the same as life originating and persisting on an exoplanet. While the 'seeding' of a planet with life on meteorites originating from a planet on which life has already evolved is plausible, this life must have had an origin on some Earth-like planet. Assuming that life on Earth originated here, there are two main contrasting hypotheses. One involves the production of organic monomers of nucleic acids and proteins in the anoxic atmosphere, energized by solar UV radiation or lightning, also ultimately energized by solar radiation. The other hypothesis, which I prefer, is that life originated at hydrothermal vents; here the energy comes from tectonic processes, yielding chemolithotrophic life. The retention of either of these mechanisms of supplying energy to life would not, however, provide an Earth-like planet orbiting a distant star that would produce a signature of life detectable remotely by reflectance spectroscopy from Earth.

There are two reasons for this lack of detectability. One is the low yield (biomass produced per unit time per unit area of planet) of either of the proposed mechanisms for the origin of life. The other is that there are no obvious spectral signatures necessarily produced by these modes of powering biota. The first problem is overcome by the evolution of the catalysed use of solar energy in photosynthesis, permitting primary production rates four or more orders of magnitude greater than those of either of the trophic modes suggested for the origin of life. The second problem is overcome by the basis of oxygenic photosynthesis with water as the source of reductant. Oxygen build-up occurs if, as happens on Earth, some of the organic matter generated in oxygenic photosynthesis escapes oxidation by heterotrophs and can be stored for millions to hundreds of millions of years by subduction through tectonic activity. Oxygen can be remotely detected, either directly or as ozone generated by UV radiation from the star. Since oxygen can be produced by abiological photodissociation of water, consideration of other characteristics of an exoplanet must be used to judge the biological component of oxygen build-up. As Lovelock pointed out almost 40 year ago, the evidence of life on an exoplanet based on the occurrence of oxygen is strengthened by finding thermodynamic disequilibrium in the atmosphere by the simultaneous occurrence of oxygen and (biogenic) methane.

The above arguments suggest not only that photosynthesis by organisms at the bacterial and algal level of organisation is the major means of producing a productive biosphere, but also that oxygen accumulation relying on the activities of these organisms provides the best means of the remote detection of life on exoplanets. While oxygenic photosynthesis on an exoplanet could fit most closely in the template of oxygenic photosynthesis on Earth if the planet was orbiting a G spectral type star like our G2V sun, oxygenic photosynthesis could occur on Earth-like planets in the continuously habitable zone of M stars (Red Dwarfs) with a much longer wavelength of peak photon absorption, beyond the approximately 1000-nm long-wavelength limit of photochemistry.

Substantial productivity by oxygenic photosynthesis under these conditions could take place using three photosystems rather than the two used on Earth.

Could such photosynthetic life be transferred between Earth-like planets? This could occur by an incoming meteorite chipping off a piece of rock bearing life from a planetary surface that, if it had escape velocity, could transfer life to another planet, at least in the same solar system. Cockell has argued that the transfer of life between planets in a solar system is relatively likely for heterotrophs that could occur quite deep (tens of mm or more) in the continental biosphere, and so would not be killed by heating of the surface of the meteorite during entry to the receiving planet's atmosphere. This means of escape from thermal sterilization would be much less likely for the necessarily surface-located photosynthetic organisms. Transfer of even heterotrophic life between solar systems is much less likely.

It is often stated that the simple organic compounds synthesised before the evolution of photosynthesis were formed in rainwater ponds rather than in the seas, possibly because the former were shallow enough for their whole water bodies to be affected by lightning energy. A more recent theory, however, has it that these organic compounds were formed in a mist, or 'organic haze', generated as gases such as N_2 mixed with the condensing water vapour, that then precipitated into the primordial (or Archaean) ocean. In parallel with organic compounds being formed in ponds or hazes some 3.5 billion years ago, Earth cooled and atmospheric water vapour condensed into rains that started to fill up the oceanic basins. (On the way to the oceans, the water also carried with it minerals, the salts of which made the seas salty.) Thus, it is thought that the organic compounds were washed from the ponds, or precipitated from the hazes by the rains, to the sea. There, in what is sometimes termed the 'primordial soup' or 'pre-biotic broth', they aggregated to form primitive cells containing more complex organic compounds surrounded by a lipid-based membrane, and developed metabolic processes and the ability to reproduce (through division) and pass down their properties to the daughter cells, all of which are characteristics of what we call life. These simple cells were thus the first life forms on our planet, and with time they developed and evolved into more complex organisms in the seas, and much, much later on land (see Figure 1.2).

It is thought that during the first 0.1 billion (100 million) years of life on Earth, the primitive cells drifting around in the early seas supplied their metabolic needs by feeding on the organic compounds washed (or precipitated) into the sea. Since, again, photosynthesis probably was absent at that time (and the organic 'foods' present in the seawater were the products of combining the primordial gases through lightning energy), these cells were, by definition, **heterotrophs**[4] (see Box 1.3). As they multiplied, a new factor came into play: competition; and as their multitude diluted their

[4] **Heterotrophs** (hetero = other; troph = feed/feeder) are those organisms that need to feed on high energy containing organic compounds for their metabolic energy needs; these are today typically represented by most of the bacteria, fungi and animals. On the other hand, **autotrophs** (auto = self; troph = feed/feeder) are those organisms that can synthesise their own food from low energy containing inorganic compounds, mainly through photosynthesis (the **photoautotrophs**, though some, the **chemoautotrophs**, use oxidation of reduced forms of elements as an energy source instead). The photoautotrophs are today typically represented by some of the bacteria (the cyanobacteria) and the plants (including algae) (see however exceptions in Box 1.3).

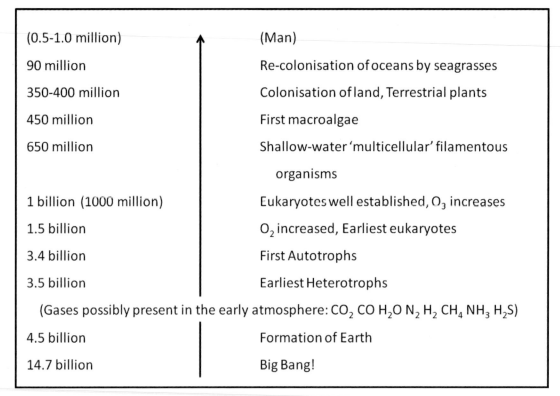

(0.5–1.0 million)	(Man)
90 million	Re-colonisation of oceans by seagrasses
350–400 million	Colonisation of land, Terrestrial plants
450 million	First macroalgae
650 million	Shallow-water 'multicellular' filamentous organisms
1 billion (1000 million)	Eukaryotes well established, O_3 increases
1.5 billion	O_2 increased, Earliest eukaryotes
3.4 billion	First Autotrophs
3.5 billion	Earliest Heterotrophs
(Gases possibly present in the early atmosphere: CO_2 CO H_2O N_2 H_2 CH_4 NH_3 H_2S)	
4.5 billion	Formation of Earth
14.7 billion	Big Bang!

Figure 1.2   A plausible time-line of life's evolution on our planet shows that, e.g. heterotrophs may have formed before autotrophs, prokaryotes (bacteria) were present for most of the time that life has existed on our planet (2–2.5 billion years), the increase in O_3 (formed from photosynthesis-derived O_2) allowed for eukaryotic filamentous algae to grow in shallow water, seagrasses re-invaded the shallow seas after their forefathers had been on land for several hundred million years and, if evolution is visualised on a 24-h clock (and after some calculations), man appeared $\frac{1}{2}$ min before midnight. Comprised by Sven Beer.

source of organic 'foods', there developed an advantage to those that could generate their own food out of inorganic chemicals. This ability evolved into what we know today as **photosynthesis**, i.e. the capability to utilise sunlight as an energy source in order to form high-energy-containing organic compounds (primarily sugars) out of low-energy inorganic compounds (largely CO_2 and water). Those later organisms were thus **autotrophs**[4], and they appeared on Earth ~100 million years after the first heterotrophic cells (i.e. ~3.4 billion years ago).

Relative to the time it took for autotrophic cells to evolve from heterotrophic ones (0.1 billion years), it took much longer (~2 billion years) for cells to evolve from being **prokaryotic**[5] to becoming **eukaryotic**[5], and this happened through the process of **endosymbiosis** (see Box 1.4). During this time the process of photosynthesis became well

[5] Karyos = nucleus and, accordingly, the **prokaroytes** lack nucleus (pro- = before) while the **eukaroytes** feature a well-defined nucleus in their cells (eu- = good/well). Today, the bacteria are prokaryotic, while all other organisms are eukaryotic. Incidentally, as we all know, the nucleus contains the hereditary basis of the eukaryotic cells and organisms in the form of linear DNA. In contrast, the prokaryotes have their DNA arranged in circular molecules.

Box 1.3 Autotrophs and heterotrophs

The term **autotroph** is usually associated with the photosynthesising plants (including algae and cyanobacteria) and **heterotroph** with animals and some other groups of organisms that need to be provided high-energy containing organic foods (e.g. the fungi and many bacteria). However, many exceptions exist: Some plants are parasitic and may be devoid of chlorophyll and, thus, lack photosynthesis altogether[6], and some animals contain chloroplasts or photosynthesising algae or cyanobacteria and may function, in part, autotrophically; some corals rely on the photosynthetic algae within their bodies to the extent that they don't have to eat at all (see Section 7.2). If some plants are heterotrophic and some animals autotrophic, what then differentiates plants from animals? It is usually said that what differs the two groups is the absence (animals) or presence (plants) of a cell wall. The cell wall is deposited outside the cell membrane in plants, and forms a type of exo-skeleton made of polysaccharides (e.g. cellulose or agar in some red algae, or silica in the case of diatoms) that renders rigidity to plant cells and to the whole plant.

[6]Most plants also contain organs that are heterotrophic (e.g. roots, which exist largely in darkness, as well as flowers and fruits, and holdfasts and sporangia in the algae, etc. However, the organism they belong to (i.e. a plant or alga) is as a whole, of course, most often autotrophic.

Box 1.4 Endosymbiosis

The theory of serial endosymbiosis is a major evolutionary step-ladder that is deeply woven into the fabric of our understanding of the origins of eukaryotic organisms. Central to this is the concept, propounded especially by Lynn Margulis, that the chloroplast and mitochondrion originated from prokaryotes that were engulfed by an ancestral phagocytotic heterotroph. The sequence of events involved is illustrated in Figure 1.3 and can be described thus (the different stages are also numbered in the figure):

1. An ancestral cell-wall-less prokaryotic cell (pale grey circle) started to develop a complex set of internal membranes around the DNA, giving rise to a primordial nucleus, and forming other organelles of the endomembrane system such as the endoplasmic reticulum, Golgi and lysosomes. At the same time, evolution of a cytoskeleton allowed this primitive cell to control flexing and in-folding of the plasma membrane.
2. This primitive phagocytotic cell (which now, with a membrane-bound nucleus, could be thought of as an ancestral eukaryote) was able to engulf bacteria (white oval with dashed border), some of which became established as endosymbionts, eventually passing on some of their DNA to the host's nucleus and becoming what we now see as mitochondria.
3. The ancestral heterotrophic eukaryote could then engulf photosynthetic bacteria (i.e. cyanobacteria; black oval in 3). Some of the latter became established in the host and, as with mitochondria, lost a large part of their DNA to the host nucleus. These endosymbionts became chloroplasts.

Although the engulfed prokaryotes gave up much of their DNA to the host nucleus, the fact that many genes are retained in a circular DNA genome within the mitochondria and chloroplasts is a strong testament to the evolutionary origins of these organelles.

4. The serial endosymbiotic theory does not stop here: What is described above is the process of primary endosymbiosis; chloroplasts initially arose when a host cell engulfed a cyanobacterium. Consequently, the chloroplasts that evolved from this process are bound by an envelope based on two membranes. Such 'primary plastids' are found in the green algae (and their descendants, the higher plants), in the red algae and in a group of algae called the glaucophytes, though whether these represent different primary endosymbiotic events or originated from a common ancestor is vigorously debated. However, many algal lines developed from further secondary endosymbiotic events where the ancestral eukaryote engulfed another alga. Plastids such as these are found in the haptophytes, cryptomonads, and in many dinoflagellates, euglenoids and stramenopiles giving rise to chloroplasts with three (e.g. euglenoids and dinoflagellates) or four (e.g. diatoms) envelope membranes (5 in Figure 1.3).

Estimates for the timing of the primary endosymbiosis suggest that this must have occurred prior to 1.6 billion years ago. The earliest likely date for the secondary endosymbiosis, in which a non-photosynthetic 'protist' captured a red algal plastid (step 4 in Figure 1.3), is ~1.3 billion years ago (see the reference of Yoon, Hackett, Ciniglia, Pinto and Battacharya, 2004), though algal groups such as the diatoms evolved much more recently (<250 million years ago).

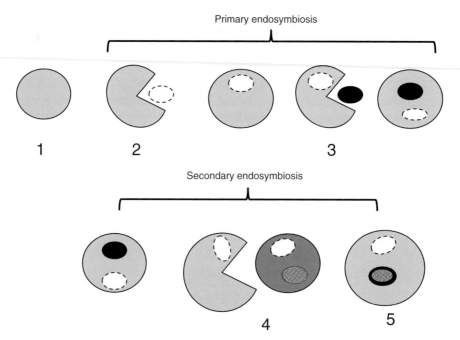

Figure 1.3   The endosymbiotic evolutionary origin of mitochondria (white oval, dashed border) **and chloroplasts** (black or hatched ovals) in algae and 'plants'. See text for details. Drawing by John Beardall.

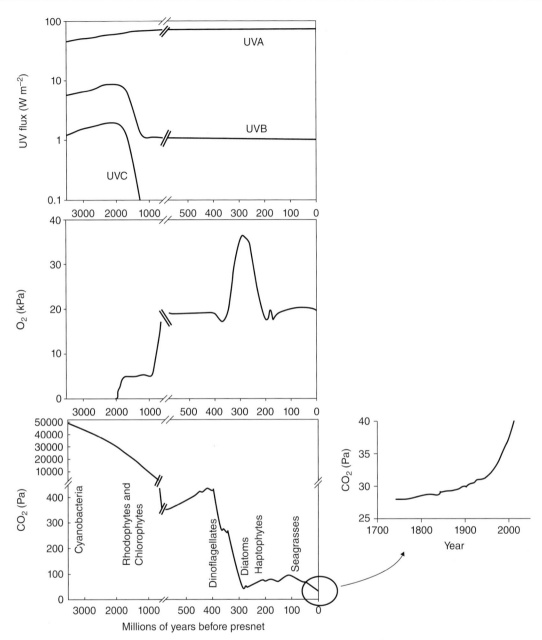

Figure 1.4   Changes, in geological time, in UV radiation (upper panel; UVA, 320–400 nm; UVB, 280–320 nm; UVC, 200–280 nm), **atmospheric oxygen** (O_2, middle panel) **and atmospheric carbon dioxide** (CO_2, lower panel); the unit Pascal (Pa) equals close to 10 ppm at standard temperature and pressure. Also indicated in the lower panel are the approximate evolutionary origins of major marine 'plant' groups. The inset shows trends in atmospheric CO_2 over the last 250 years. Adapted from: Beardall J, Raven JA, 2004. The potential effects of global climate change on microalgal photosynthesis, growth and ecology. *Phycologia* 43: 26–40. Copyright (2013), with permission from Allen Press Publishing Services.

established as the ultimate way of forming high-energy organic compounds, as did that of respiration, the most efficient form of which relied on a by-product of photosynthesis: molecular oxygen (O_2); this type of respiration is also called oxidative respiration. As photosynthesising organisms multiplied and evolved, O_2 levels slowly rose in the atmosphere, transforming it from a reducing to an oxidising one. While such an atmosphere indeed is favourable for oxidative respiration, it inhibits photosynthesis of most terrestrial plants (but more about that in Chapter 6).

This is not a book on evolution, so that subject is dealt with only briefly as it pertains to the evolution of marine photosynthetic organisms. During the first 2 billion years of organisms' evolution on the planet, levels of O_2 derived from photosynthesis rose in the oceans as well as in the ocean-equilibrated atmosphere (again, see Box 1.1). This led to the formation and successive accumulation of ozone (O_3) in the upper atmosphere, which protected both the prokaryotic (which ruled the seas during 2 billion years) and eukaryotic (which became well established ~1 billion years ago) cells from part of the harmful UV radiation present in the upper water layers. Consequently, the unicellular organisms could move up in the water column towards surface waters, the higher irradiances of which further favoured photosynthesis (especially of eukaryotic phytoplankton, which generally utilise higher irradiances than cyanobacteria, see Section 9.3). At the same time, competition to obtain both organic (for the heterotrophs) and inorganic (for the autotrophs) nutrients increased as the cells multiplied and resources, accordingly, became scarcer. For the autotrophs, this meant that there was an advantage if they could live close to the shores where inorganic nutrient concentrations were higher (because of mineral-rich runoffs from land) than in the upper water layer of off-shore locations. However, living closer to shore also meant greater effects of wave action, which would alter, e.g. the light availability on shorter than diel-term frequencies. Under such conditions, there would be an advantage to be able to stay put in the seawater, and under those conditions it is thought that filamentous photosynthetic organisms were formed from autotrophic cells (*ca.* 650 million years ago), which eventually resulted in macroalgae (some 450 million years ago) featuring holdfast tissues that could adhere them to rocky substrates. (While most of today's 'macroalgae' are eukaryotic, there are also cyanobacterial filaments that stick to bottom substrates, see Section 2.1.)

Very briefly now, the green macroalgae were the ancestors of terrestrial plants, which started to invade land *ca.* 400 million years ago (followed by the animals). It is interesting (and logical) to note that the successful plant invasion of land caused an increased rate at which O_2 levels rose in Earth's atmosphere (see Figure 1.4). Long after terrestrial plants were established, there was an interesting (especially to SB and MB) secondary re-invasion of some grasses back to the marine environment which (luckily for us that study them), some 90–100 million years ago, gave rise to the seagrasses (see Section 2.5).

Chapter 2

The different groups of marine plants

There are tens of thousands of marine 'plant' species (see our broad definition of plants in the introduction to this part), and new ones are being discovered each year (see, however, Box 2.4 regarding reductions of some macroalgal genera and seagrass species). Some are prokaryotic, and while they are sometimes still called 'bluegreen algae' (because of their sometimes blue-green appearance), a more appropriate name is cyanobacteria (since they in fact are bacteria, albeit photosynthetic ones). They will be described in the following section, and their photosynthetic properties will be treated in Section 7.1. Among the eukaryotic marine plants, we differentiate between algae and seagrasses. The algae can, in turn, be divided into microalgae (Sections 2.2 and 7.1 for a description, and photosynthetic properties, respectively) and macroalgae (Sections 2.4 and 7.3, accordingly) as determined roughly by them being observable through a microscope or with the naked (or spectacled) eye, respectively. The seagrasses, or marine angiosperms, which are also macroscopic, differ from the macroalgae in that they are angiosperms, i.e. flowering plants with characteristics similar to those of terrestrial grasses (except that they can complete their entire life cycle whilst being submerged in seawater). They, and their photosynthetic traits, will be described in Sections 2.5 and 7.4, respectively. Finally, the cyanobacteria and algae acting as photosynthetic symbionts (also called photosymbionts) in many marine invertebrates will be described in Section 2.3, and what little we know of their photosynthetic properties will be treated in Section 7.2. We were deliberating whether or not to include the mangroves[1] in the marine plant realm, but

[1]Mangroves are those trees that have their roots submerged in seawater. We (SB and MB) have, however, observed that some mangroves in areas where the tidal amplitude is high can be totally submerged for extended periods of time, and that they photosynthesise also when the leaves and, indeed, entire trees are under water. Therefore, one could include those trees as being part of the marine plants. Mangroves do not, however, reproduce under water, and so we leave their inclusion into the marine plant realm open for the time being.

Photosynthesis in the Marine Environment, First Edition. Sven Beer, Mats Björk and John Beardall.
© 2014 John Wiley & Sons, Ltd. Published 2014 by John Wiley & Sons, Ltd.
Companion Website: www.wiley.com/go/beer/photosynthesis

decided against it at this stage; perhaps we will do so in coming editions of this book.

2.1   Cyanobacteria

A large part of the phytoplankton of the seas consists of prokaryotic unicellular or filamentous organisms known as cyanobacteria. The name stems from them often appearing as bluish (cyano = blue), or blue-green, and they were also formerly called blue-green algae. The latter name is, however, less used today since 'algae' refers to eukaryotic organisms. (Also, the 'phyto' (= plant) part of 'phytoplankton' indicates eukaryotic organisms, but we will here treat most cyanobacteria as part of the phytoplankton.) Unlike the more commonly known bacteria, cyanobacteria of course carry out oxygenic[2] photosynthesis.

As stated in the previous chapter, cyanobacteria were, some 3.4 billion years ago, the first photosynthetic (and thus autotrophic) organisms to evolve in the oceans. Cyanobacteria are found in a very wide range of environments, from freshwaters to high salinity systems and from Polar Regions to hot springs; there are some 1500 species of cyanobacteria recognised today, of which ~10% are marine. In marine systems, unicellular species such as *Synechococcus* and *Prochlorococcus* are extremely important contributors to phytoplankton productivity, and in tropical waters the colonial cyanobacterium *Trichodesmium* can form extensive blooms. Not all marine cyanobacteria are planktonic, however, and they can be significant contributors to the microphytobenthos (microscopic primary producers found attached to, or between, particles in sediments or as mats attached to surfaces); some are macroscopic and resemble filamentous macroalgae (see Figure 2.1a), while *Prochloron* is found in symbiotic association with ascidians in tropical waters; many other such photosymbiotic associations with marine invertebrates exist among the cyanobacteria (see Section 2.3).

While cyanobacteria as a whole are quite varied and fall into 5 major groups, the marine cyanobacteria fall into three broad taxonomic categories: The **unicellular cyanobacteria** include the important species *Synechococcus* and *Prochlorococcus*. Previously, because of their small size, these taxa (*Prochlorococcus* especially) had been largely overlooked. With recent advances in collection and detection (including molecular) techniques, however, it is now recognised that these organisms are a major component of the picoplankton (Table 2.1) and thus contribute significantly to the primary production of the oceans. Most unicellular cyanobacteria are unable to assimilate N_2 gas[3], but some larger forms such as *Crocosphaera* (with diameters up to 7 μm) and *Gloeothece* have this capability. The second group comprises the **filamentous cyanobacteria that do not form heterocysts**[3], such as *Oscillatoria*, *Lyngbya* and *Phormidium*. Many of these species are important in estuarine and benthic habitats. Although some species of *Oscillatoria* and *Lyngbya* will fix N_2 under darkness or anaerobic conditions, they are not major contributors to nitrogen fixation. The filamentous cyanobacterium *Trichodesmium* is,

---

[2] Oxygenic photosynthesis involves the production of O_2 as electrons are withdrawn from water in photosystem II (see Chapter 5). This contrasts with non-oxygenic photosynthesis where, for example, electrons are withdrawn from hydrogen sulphide, as in some photosynthetic bacteria.

[3] Some cyanobacteria have the capability to assimilate N_2 dissolved in seawater and transform it to ammonia (NH_3), the latter of which is used for forming amino acids and, ultimately, proteins. In the filamentous cyanobacteria that carry this capacity, the N_2 fixation and eventual reduction usually occur in specialised cells, the heterocysts, but the recognized number of exceptions to this is growing. See further in Box 2.1.

Figure 2.1   Marine cyanobacteria include mostly microscopic planktonic forms such as depicted in Figure I.1(a), but can also be macroscopic and benthic such as *Symploca hydnoides* (a). *Trichodesmium* is a marine filamentous, N_2-fixing, cyanobacterium that often forms spherical colonies (left in b). Picture (c) depicts a mixture of cyanobacteria in a cyanobacterial bloom of the Baltic Sea; the dominating genus *Nodularia* forms loosely arranged rounded colonies of 'curly' filaments, while *Aphanizomenon* arrange their straight filaments into parallel colonies. Photos by Mats Björk (a) and Birgitta Bergman (b and c).

however, an exception and is an extremely important N_2 fixer in tropical and sub-tropical oceans, where it is often found in vast surface blooms (see the box by Birgitta Bergman). The third group is the **heterocystous filamentous cyanobacteria**. This group includes the important toxic species of *Nodularia*, *Anabaena* and *Aphanizomenon*, which are common bloom formers in brackish waters such as the Baltic Sea or coastal lagoons such as the Gippsland Lakes in southern Australia.

2.2   Eukaryotic microalgae

There is a range of eukaryotic algae that are found as phytoplankton in the oceans, with some 20 000 species currently recognised. These organisms evolved some 1.5 billion years ago, after the cyanobacteria had ruled the seas as the sole oxygenic (= oxygen-evolving) photosynthesisers for almost 2 billion years (see the previous chapter). The major eukaryotic contributors to the phytoplankton are the

Box 2.1   N₂ Fixation by Marine Cyanobacteria

by **Birgitta Bergman**, Department of Ecology, Environment and Plant Sciences, Stockholm University, SE (birgitta.bergman@su.se)

The conversion of the plentiful atmospheric N_2 gas ($\sim$78% in air) into bio-available N-rich cellular constituents is a fundamental process that sustains life on Earth. For unknown reasons this process is restricted to selected representatives among the prokaryotes: archaea and bacteria. N_2 fixing organisms, also termed diazotrophs (dia = two; azo = nitrogen), are globally wide-spread in terrestrial and aquatic environments, from polar regions to hot deserts, although their abundance varies widely. However, the existence of diazotrophs in marine environments was overlooked until in recent decades. New efficient tools for monitoring microbes in oceans (ocean-going cruises and satellite monitoring), combined with the recent use of large-scale environmental sequencing of marine microbes (metagenomic analyses), have however dramatically reshaped our understanding of the occurrence and physiological performances of marine diazotrophs. Today it is estimated that diazotrophs deliver about 0.15–0.20 Gt 'new' N per year into the marine biosphere, and their importance in the global biogeochemical cycle of N is thereby undisputable. Cyanobacteria dominate marine diazotrophs and occupy large segments of marine open waters, where widespread and massive cyanobacterial surface blooms may appear, and cyanobacteria are also known to establish biofilms and microbial mats in coastal areas.

The photosynthetic and highly pigmented (by chlorophyll and phycobilins, see Section 4.3) character of cyanobacteria gives these organisms great competitive advantages in oligotrophic (nutrient-deprived) ocean waters. The full energy demand is derived from light (via ATP formed by photo-phosphorylation, see Section 5.3) and the carbon via CO_2-fixation and -reduction (see Chapter 6), two pivotal metabolites required for sustained N_2 fixation, which is a highly energy-demanding process. As in all diazotrophs, the nitrogenase enzyme complex (encoded by *nif*K, *nif*D and *nif*H genes) of marine cyanobacteria requires high Fe levels (a cofactor in the NifKD/Mo-Fe, and the NifH/Fe, proteins). Another key nutrient is phosphorus, received from P-rich upwelling waters and aeolian dust depositions, which has a great impact on growth and N_2 fixation in marine cyanobacteria.

Traditionally, the most widely used method for monitoring nitrogen fixation is the 'acetylene reduction assay', an indirect method in which nitrogenase catalyses the conversion of acetylene gas (C_2H_2) to ethylene gas (C_2H_4), subsequently quantified using gas chromatography and, more recently, photodetectors. However, N_2 fixation may be assayed directly by measuring the fixation of a $^{15}N_2$ tracer gas into cells using mass spectroscopy. Recent data suggest that the 'traditional' way of adding $^{15}N_2$ gas as a 'bubble' into the incubation bottle, compared to dissolving the $^{15}N_2$ gas in the sea water prior to the addition of organisms/cells, may underestimate the amount of N_2 fixed 2- to 6-fold. This implies a currently underestimated contribution of cyanobacterial N_2 fixation in marine environments. The *nifH* gene, *nifH* gene transcript and NifH protein abundance are today also frequently used as means to monitor the relative distribution of marine diazotrophic cyanobacteria and their diazotrophic capacity, although direct measurements are prerequisites to obtain proper N_2 fixation activities.

Initially, the focus on marine N_2 fixation was on the planktonic tropical and subtropical diazotrophic genus *Trichodesmium*, which is widespread in all oceans at higher temperatures (>20 °C) where it forms highly buoyant colonies (see Figure 2.1b). *Trichodesmium* belongs to Section III filamentous cyanobacteria, lacking the well-known nitrogenase enzyme protective cell type, the heterocysts. Section I-III cyanobacteria all lack heterocysts and are therefore forced to fix N_2 at night (i.e. in darkness) to protect the oxygen-sensitive nitrogenase from O_2 generated in their own photosynthesis. An exception is *Trichodesmium*, which fixes N_2 in full daylight and in oxygenated surface waters. *Trichodesmium* differentiates yet another specific cell type, diazocytes, in which the nitrogenase enzyme is synthesized and potentially protected. Like heterocysts, the diazocytes differentiate as a response to combined nitrogen deprivation. With this strategy, they can directly use the photosynthetically derived energy for N_2 fixation and do not have to rely on stored substances (e.g. glycogen) as do other Section III cyanobacteria. Such capacities may explain its efficient N_2 fixation and ubiquity in oceans and give this genus a cornerstone role in the overall global nitrogen economy. For still unknown reasons, heterocystous cyanobacteria are rare in fully marine open waters while some genera form massive surface blooms in lower-salinity waters (e.g. the brackish Baltic Sea). Heterocystous forms, such as the genus *Calothrix*, are also distinct components of coastal shallow-water cyanobacterial mats where they actively fix N_2.

Today, unicellular cyanobacteria attract legitimized increased interest as N_2 fixers in marine ecosystems. Recent model-based estimates of N_2 fixation suggest that unicellular cyanobacteria (Section I-II) indeed contribute significantly to global ocean N budgets. Being oxygenic phototrophs, their N_2 fixation by necessity takes place at night and a daily *de novo* biosynthesis of *nif* gene transcripts and Nif proteins, and their subsequent degradation, are required. Representatives of this unicellular category include planktonic genera such as *Chrocosphaera* (2.5–6 µm diameter) and *Cyanothece* (varying cell diameter). *Chrocosphaera spp.* may reach cell densities as high as 1000 cells/ml in the euphotic zone (see Section 3.1) of oligotrophic waters (Atlantic and Pacific oceans). The genome of *Chrocosphaera watsonii* (WHOI8501) has been sequenced (6.24 Mbp), as has the genome of the type strain *Cyanothece* sp. ATCC 51142 (4–5 µm diameter; genome size 5.46 Mbp), as well as an additional six *Cyanothece* genomes. The genus *Cyanothece* is a morphologically and ecologically versatile group and contributes significantly to marine nitrogen budgets. The genomes verify the genotypic and phenotypic flexibility (e.g. varied cell sizes and shapes; numerous transposons) and suggest a vivid loss and gain of genetic material over evolutionary time. However, a stable cyanobacterial core is apparent that underpins their anaerobic night-time N_2 fixing physiology. The fact that the *Cyanothece* strain forms a single clade together with strains of *Crocosphaera* and yet another marine unicellular cyanobacterium, UCYN-A, suggests their common ancestry. The recently discovered and hitherto uncultivated UCYN-A potentially evolved from a *Cyanothece*-like ancestor via genome reduction. The minute and streamlined genome (1.44 Mbp) of UCYN-A indeed lacks several key genes, for instance genes encoding the oxygen-generating photosystem II (see Chapter 5), Rubisco (see Chapter 6) as well as the tricarboxylic acid cycle. This implies a photoheterotrophic lifestyle and dependence on a photosynthetic host, now identified as a prymnesiophyte. UCYN-A receives photosynthate from the host while it delivers fixed nitrogen in return. UCYN-A shows a wide distribution including cooler marine waters, and more recently a UCYN-B type was described.

Besides the symbiotic UCYN-A strains, the oceans also hold a number of other understudied diazotrophic cyanobacteria that live in symbiosis with eukaryotic hosts. For instance, cyanobacteria colonize marine diatoms that form large surface blooms in, e.g. the Indian and the tropical Atlantic oceans (outside the Amazon River plume). These cyanobacteria are of the heterocystous type (Section IV) and are closely related to well-known heterocystous genera (*Anabaena/Nostoc*), although they form a separate clade in phylogenetic trees. Some cyanobacteria are intracellular, such as *Richelia* in the diatoms *Rhizosolenia* and *Hemialus*, and some are extra-cellular (on the frustule), such as *Calothrix* on the diatom *Chaetoceros*. The nitrogen fixed by the cyanobacterium is transferred to the host, while carbon compounds move in the opposite direction. Yet other unicellular cyanobacteria, some potential diazotrophs, are known to colonize dinoflagellates, radiolarians and tintinnids, but our understanding of these are still rudimentary in particular with regard to their diazotrophic significance in marine waters.

Other marine diazotrophic, but largely neglected, populations are those occupying coastal habitats, such as sandy and muddy bottoms and rocky shores, which are almost invariably covered by microbial mats containing cyanobacteria. Particularly thick cyanobacterial mats are found in the regularly flooded mangrove forests worldwide. Yet other overlooked diazotrophic cyanobacterial populations are those that occur as epiphytes on, e.g. seagrasses, macroalgae and corals, the role of which as contributors to the N budget of their 'substrate' organism may be significant but still unknown. There is room for much needed research here!

diatoms and dinoflagellates, though the haptophytes (especially the coccolithophores) can form enormous blooms in the ocean. In contrast, there are only a few green algae and even fewer red algal species that are found in the marine plankton. For convenience, we often divide the phytoplankton into different size classes, the pico-phytoplankton (0.2–2 μm effective cell diameter, ECD[4]); the nano-phytoplankton (2–20 μm ECD) and the micro-phytoplankton (20–200 μm ECD). The major algal groups found in the different size classes are shown in Table 2.1, and some examples are depicted in Figure 2.2. From Table 2.1, it can be seen that most of the major marine microalgal groups are found in all three size classes, e.g. it is possible to find diatoms and

dinoflagellates that fit in all three classes. Uitz, and co-workers in a 2010 paper estimate that these plants utilise 46 Gt carbon yearly, which can be divided into 15 Gt for the microphytoplankton, 20 Gt for the nanophytoplankton and 11 Gt for the picophytoplankton. Thus, the very small (nano- + pico-forms) of phytoplankton (including cyanobacterial forms) contribute 2/3 of the overall planktonic production (which, again, constitutes about half of the global production, but we leave some room for the macrophytes too).

The major characteristics of the phytoplankton groups commonly found in the oceans are presented in Table 2.2. (The characteristics of the marine cyanobacteria have been described above.) Of the eukaryotic phytoplankton, three groups are especially relevant: The **diatoms** are perhaps both one of the best-known groups of eukaryotic phytoplankton and the most important in terms of their contribution (~40%) to oceanic primary productivity. One of their

---

[4]Efective cell diameter, or ECD, is a practical means to estimate the diameter of a phytoplanter that is not perfectly round but, as most species, is of odd shapes; their ECD is thus the diameter of an equivalent sphere.

Table 2.1 Distribution of different size classes among the major groups of marine microalgae and cyanobacteria. Note that all groups are represented in the nanoplankton, whereas larger and smaller forms are somewhat more restricted in their taxonomic distribution (see text for details).

Size class	Picoplankton	Nanoplankton	Microplankton
Cyanobacteria	✓	✓	✓*
Diatoms	✓	✓	✓
Dinoflagellates	✓	✓	✓
Green algae	✓	✓	✓
Euglenids (a type of green alga)			✓
Prasinophyceae (a type of green alga)	✓	✓	✓
Pelagophyceae	✓	✓	
Haptophytes		✓	

*Colonial forms such as *Trichodesmium* can be larger.

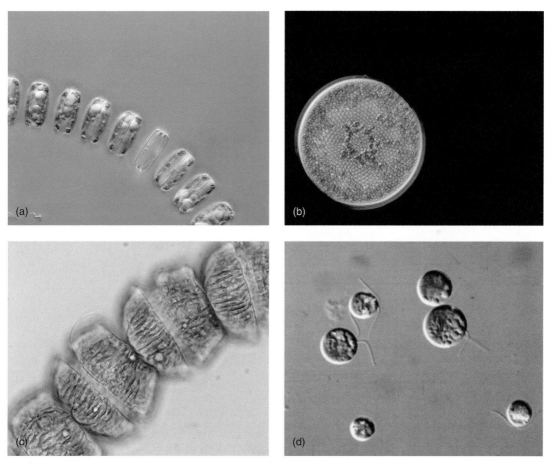

Figure 2.2 Some representatives of **marine eukaryotic microalgae.** (a) *Thalassiosira* sp., a small, chain-forming diatom typical of open ocean environments; depending on species, *Thalassiosira* cells can be 5–50 μm in diameter. (b) A large marine diatom, *Coscinodiscus* sp.; some species of this genus can be up to 500 μm across and are then visible to the naked eye. (c) A toxic dinoflagellate, *Gymnodinium catenatum*, common in coastal environments; cell diameters usually range from 30–45 μm. (d) A green microalga, *Dunaliella tertiolecta*, ~10 μm in diameter. Photos by Martina Doblin (a–c) and John Beardall (d).

Table 2.2 Some characteristics of the major groups of marine phytoplankton.

Group	General	Motility	Main light-harvesting pigments	Cell wall	Storage compound
Cyanobacteria	unicellular, filaments, some colonial	some species glide	Chl a, phycobilins (β-carotene, myxoxanthin)	peptidoglycan and lipopolysaccharide	cyanophycean starch
Dinoflagellates	mostly unicellular	two: one longitudinal, one in a furrow	Chl a, c, phycobilins or fucoxanthin	naked or covered in cellulosic plates (thecae)	starch, some lipids
Green algae	unicellular, some filamentous or colonial	smooth on motile species, some non-flagellate	Chl a, b, β-carotene	flagellate cells have a glycoprotein coat, non-flagellate cells have polysaccharide (cellulose) walls	starch, lipids
Prasinophyceae (a subclass of green algae)	unicellular	1–4 smooth flagella on vegetative cells	Chl a, b, β-carotene	organic scales, *Micromonas* naked	starch
Pelagophyceae	unicellular	laterally inserted on zoids; one long with hairs, one short smooth	Chl a, b, β-carotene, butanoyloxy-fucoxanthin	cellulose	chrysolaminarin
Prymnesiophyceae	unicellular	two smooth flagella plus a haptonema	Chl a, c_1, c_2, β-carotene, 19'-hexanoyloxyfucoxanthin	some may have scales, e.g. calcite in coccolithophorids	chrysolaminarin
Diatoms	unicellular, filamentous, small colonies	Some glide, some have single lateral flagellum	Chl a, b, β-carotene, fucoxanthin	silica frustule	chrysolaminarin, lipids

highly characteristic features is the possession of a **silica cell wall**, the frustule. This structure is highly ornamented in a way that is species specific, and this is often therefore used as a means of identification[5].

[5]The phytoplankton have classically been identified microscopically by morphological features such as the shape and ornamentation of the frustules (for diatoms). Today, however, identification based on a sequence of their DNA is taking over from the microscopic observation method. This is both more exact and lucky since plankton taxonomists are becoming a very rare species.

The **dinoflagellates** possess two flagella and can form significant blooms (red tides), which are often toxic. Finally, the **Prymnesiophyceae** (a subtaxon class of the group known as Haptophytes), are characterised by having 2 flagella and a fine whip-like structure called the haptonema. Prymnesiophyte cells are covered in scales and in the most important group, the **coccolithophores**, these scales (called coccoliths) are calcified (see, e.g. Figure 2.5a); there is a wide array of species-specific coccolith architecture in coccolithophores. Several prymnesiophyte genera such as *Emiliania* and the non-coccolithophore *Phaeocystis* will form extensive blooms in the oceans and some other species such as *Chrysocromulina* are toxic.

Because of their ecological importance, diatom genera such as *Thalassiosira* and the coccolithophore *Emiliania* have been subject to intense investigations of their photosynthetic physiology. On the other hand, historically, species studied in the laboratory (such as the diatom *Phaeodactylum tricornutum* and the green alga *Dunaliella tertiolecta*) were chosen on the basis of their robustness in culture rather than their significance in nature.

Many marine microalgae have been utilised for the production of useful compounds. This is a field of increasing interest, and will only be described briefly in Section 12.1; the interested reader is referred to, for example, Borowitzka and Hallegraeff (2007). Very briefly here, the species used are mainly *Dunaliella* for β-carotenes, *Haematococcus* for the carotenoid astaxanthin and for Ω- (omega)-fatty acids, and *Nannochloropsis* for various nutritional additives and biofuels (although the economic feasibility of the latter is very questionable).

2.3   Photosymbionts

Many primarily non-photosynthetic organisms have developed symbioses with microalgae and cyanobacteria; these photosynthetic intruders are here referred to as **photosymbionts**. This association is a practical way to capture sunlight and provide the hosts with energy under otherwise often nutrition-poor conditions. Most photosymbionts are endosymbiotic (living within the host), and if they are microalgae they belong almost exclusively to the dinoflagellate genus **Symbiodinium**, also commonly known as **zooxanthellae**. In almost all cases, these micro-algae are in symbiosis with invertebrates. Here the alga provides the

animal with organic products of photosynthesis, while the invertebrate host can supply CO_2 and other inorganic nutrients including nitrogen and phosphorus to the alga (see, however, Section 7.2 on uncertainties regarding the CO_2 supply). In cases where cyanobacteria form the photosymbiont, their 'caloric' nutritional value is more questionable, and they may instead produce toxins that deter other animals from eating the host, or provide the host with UV-protecting amino acids, etc.

Many reef-building (see, e.g. Figures I.1c and 2.3a), as well as soft, **corals** contain symbiotic zooxanthellae within the digestive cavity of their polyps, and in general corals that have symbiotic algae grow much faster than those without them. The dependence of the host on the photosymbiont may be so essential that many coral species hosting zooxanthellae cannot grow deeper than the extension of their photic zone (i.e. some 100 m in clear waters, see Section 7.2 for an account of photosynthesis in corals). The loss of zooxanthellae from the host is known as coral bleaching, which has been associated with global warming of the seas and that can lead to the deterioration of reefs. Not only has coral bleaching been a huge source of funding for scientists researching it, but, fortunately (or unfortunately for some of those scientists), the re-acquisition of other types (or clades) of *Symbiodinium*, possibly better adapted to high temperature, has now initiated the re-building of those deteriorating reefs in many locations worldwide. In addition to providing the host with organic nutrients and O_2 during the daylight hours, the photosynthetic activity of the zooxanthellae is also known to stimulate calcification. Regarding some photosynthetic qualities of zooxanthellae in reef-building corals, see Section 7.2.

Out of the many organisms, other than corals, that harbour photosymbionts can be mentioned some **jellyfish** carrying symbiotic

Figure 2.3   Photosymbiont-containing invertebrates. Depicted are the staghorn coral (*Acropora*, a) and the giant clam *Tridacna* (b), both containing zooxanthellae, and a cyanobacterial-containing sponge (c). The photosynthetic activity of such invertebrates can easily be measured *in situ* by pulse amplitude modulated (PAM) fluorometry (d, see Section 8.3 for a description of the technique; depicted is SB with a 'Diving-PAM'). Photos by Katrin Österlund.

algae in their tentacles. Famously, *Cassiopeia xamachana* lies upside down on the sea floor of shallow tropical waters, allowing the zooxanthellae to receive maximum daylight for photosynthesis. Similarly, some **sea anemones** supply part of their nutritional demand by the photosynthetic activity of photosymbiotic zooxanthellae and zoochlorellae (a form unicellular green algae). The sea anemone benefits by receiving O_2 and food in the form of, e.g. glycerol, glucose and alanine, all stemming from the photosynthetic activity of the photosymbiont. Many **mussels**, e.g. the giant clam *Tridacna* (Figure 2.3b) have a symbiotic pop-

ulation of photosynthetic zooxanthellae. The algae provide the clams with supplementary nutrition, which may aid them to grow large even in nutrition-poor waters such as found in and around many coral reefs. Also, a number of **sarcodines** (a subgroup of protozoa including the amoebas), as well as Heliozoa, Radiozoa and Foraminifera, have symbiotic algae living within their cells. Regarding the foraminifera, it has been shown that the photosynthetic activity of their various photosymbionts also enhances their rates of calcification.

Certain **sea slugs** contain functional chloroplasts that were ingested (but not digested) as

Box 2.2   Photosymbiosis in sponges

Laura Steindler, Haifa University, IL (lsteindler@univ.haifa.ac.il)

Sponges are aquatic sessile invertebrates feeding on bacteria and other small particles by means of water filtration. They are found in association with a large variety of heterotrophic bacteria and photosynthetic organism. The latter include cyanobacteria, dinoflagellates, rhodophytes, chlorophytes and diatoms. These photosymbionts provide an additional source of nutriment (photosynthates) and, possibly, fixed nitrogen to the sponge host. Thus, it is believed that through translocation of fixed carbon the photosynthetic symbionts could increase the overall holobiont fitness (holobiont = host+(photo)symbiont). Photosymbiosis is a common feature in marine sponges and accounts for 1/3–3/4 of coral reef sponges in tropical regions and over half of sponges from temperate ecosystems. The most prevalent photosymbionts appear to be cyanobacteria including multiple phylogenetic lineages and species related to *Oscillatoria spongeliae*, *Synechocystis trididemni* and *Candidatus* Synechococcus spongiarum. The latter is the most commonly reported and widespread sponge-cyanobacterial symbiont found in taxonomically diverse hosts from geographically distant regions. *Candidatus* Synechococcus spongiarum is a single-celled cyanobacterium that occurs in the external (ectosomal) regions of the sponge body and has never been isolated in culture. The acquisition of this symbiont was suggested to be 'vertical' both because these cyanobacteria were observed within the sponge eggs, sperm and larvae, as well as because of its absence in environmental seawater samples, as deduced from 16S rRNA clone libraries. Nevertheless, recent advances in high-throughput sequencing methods have enabled the detection of cyanobacterial (and other non-photosynthetic bacterial) sponge symbionts at very low concentrations in seawater samples, suggesting that members of the rare seawater biosphere may serve as seed organisms for widely occurring symbiont populations in sponges.

Microalgae of the genus *Symbiodinium* (or zooxanthellae), which are better known as providers of energy in coral hosts, are also found as symbiotic partners within the sponge holobiont. Although zooxanthellae are found in a variety of sponges, the symbiosis is most common in sponges of the family Clionidae (bio-eroding sponges), where boring activity and growth rate appear to be enhanced through the photosynthetic activity of the symbionts. Zooxanthellae in sponges are found intracellularly, mainly in the ectosome where light levels are highest. We still have only a very limited understanding on the metabolic integration between zooxanthellae and their sponge host. Although it is often hypothesised that zooxanthellae benefit from the association with the host by trading their excess photosynthate for nitrogen, a study that utilised stable isotopic labels provided evidence only for transfer of carbon from alga to host, while no clear transfer of nitrogenous compounds from host to algae was evident, although the latter results remained equivocal. On the other hand, heterotrophic microbial symbionts, including bacteria from kingdom Archaea, are receiving increasing attention as they may play key roles in the nitrogen metabolism of sponges. Another important theme investigated in symbiosis research is the specificity of the interaction between the players. Different zooxanthellar clades have been identified by molecular techniques. Some clionaid sponges do not host zooxanthellae, while within those that do, some form multi-cladal partnerships, and others appear to host a single zooxanthellar 'partner', raising the questions on how specificity between sponge and zooxanthellae is achieved and what are the advantages of hosting each of the different clades. The number of studies examining specificity of zooxanthellae in different sponges is, however, still too low to reach ecologically relevant conclusions.

The molecular mechanisms involved in the initiation, establishment and maintenance of sponge photosymbiosis have been explored in the Mediterranean sponge *Petrosia ficiformis*, where cyanobacteria are found as facultative symbionts: i.e. specimens in the light harbour cyanobacteria while specimens in dark caves do not contain them. Transcriptional profiles of specimens with and without cyanobacteria revealed several differences in genetic expression, suggesting that some of these genes may function in the recognition of the symbionts or inactivation of an immune response enabling discrimination of true symbionts from potential pathogenic microorganisms.

Future research directions will certainly take advantage of the novel high-throughput sequencing methods. Research on sponge symbiosis has entered the '-omics' era with metagenomes and single-cell genomics utilised to reveal interesting metabolic features of microbial symbionts. Genome sequencing of yet uncultured cyanobacterial sponge photosymbionts (e.g. *Candidatus* Synechococcus spongiarum) could reveal adaptations of early free-living *Synechococcus* species to life in symbiosis. Parallel efforts are required to implement novel techniques of isolation and growth of sponge photosymbionts, which will be crucial to test hypotheses provided by culture-independent studies, to determine the physiology of photosymbionts and reveal their function within the sponge host.

Photosymbioses in sponges have not been studied much, but here are a few references for those interested in pursuing such research in the future: Usher KM, Fromont J, Sutton DC, Toze S, 2004. The biogeography and phylogeny of unicellular cyanobacterial symbionts in sponges from Australia and the Mediterranean. Microbial Ecology 48: 167–177; Weisz JB, Massaro AJ, Ramsby BD, Hill MS, 2010. Zooxanthellar symbionts shape host sponge trophic status through translocation of carbon. Biological Bulletin 219: 189–197; Steindler L, Schuster S, Ilan M, Avni A, Cerrano C, et al., 2007. Differential gene expression in a marine sponge in relation to its symbiotic state. Marine Biotechnology (NY) 9: 543–549.

part of larger algae such as *Ulva* or coenocytic green algae (see below). After digesting the rest of the alga, these chloroplasts are imbedded within the slugs' digestive tract in a process called kleptoplasty (the 'stealing' of plastids). Even though this is not a true symbiosis (the chloroplasts are not organisms and do not gain anything from the association), the photosynthetic activity aids in the nutrition of the slugs for up to several months, thus either complementing their nutrition or carrying them through periods when food is scarce or absent.

Second to corals, the perhaps best-known photosymbiotic association with invertebrates is that with **sea sponges** (see, e.g. Figure 2.3c). These animals carry a variety of photosymbionts. In some, such as the breadcrumb sponge (*Halichondria panicea*), green algae live close to their surface. The alga is thus protected from predators, while the sponge is provided with oxygen and sugars, which can account for 50–80% of its nutritional needs. In others, and more commonly, either cyanobacteria or zooxanthellae constitute the photosymbiont, but it is unclear how much nutrition they can provide the sponge host. See further in Box 2.2 about photosymbiosis in sponges.

2.4   Macroalgae

The **macroalgae** (also called **seaweeds**[6]) are, as their name suggests, macroscopic, i.e. they can easily be seen with the naked eye. (Since some cyanobacteria that are attached to substrates are macroscopic too, those that are so could also be functionally included in the macroalgae.) The *ca.* 20 000 species of eukaryotic macroalgae can be divided into 3 groups (or Divisions): the green, the brown and the red algae (see examples in Figure 2.4). The body of a macroalga is called the thallus (from the Greek thallos = sprout/twig; thalli in plural), and it contains in most cases, in comparison with terrestrial plants, rather undifferentiated tissues. There are, however, many exceptions: Most macroalgae have holdfasts', which comprise a specialised, sometimes rootlike, tissue that anchors their thalli to the rocks on which they grow. The cells of holdfasts secrete a protein-based adhesive, which has also attracted the interest of applied scientists because of its extreme strength. (Also, being a 'natural product', this compound has triggered pharmaceutical start-ups with the aim of supplying a 'natural glue' for surgical application.) Tissue differentiation also occurs in the middle parts of thick thalli, which, e.g. do not contain chloroplasts. Also, the cells and tissues involved in the propagation of algae are differentiated from those that make up their bulk.

---

[6]The **macroalgae** have been referred to by different names: Since they are mostly benthic (living close to the bottom), **phytobenthos**, or more correctly **macrophytobenthos** can be an appropriate word describing them (as well as the forthcoming seagrasses) with respect to their growth habitat. The word **seaweed** associates with their often 'weedy' appearance (e.g. as sometimes disturbing swimmers and beach-visitors), but since the word 'weed' is associated with terrestrial nuisance plants, the seaweeds should not be mixed up with seagrasses, the latter of which indeed are marine higher (flowering, like terrestrial grasses) plants (see later in this chapter).

Another interesting exception to the generally considered 'undifferentiated' thalli of algae can be found in some kelps (belonging to the brown algae), in which vascular bundles are present, which improves the ability to transport materials, especially photosynthate, along the large thalli of such algae. In this section, we will only give a short description of the different macroalgal groups as a background to the forthcoming chapters; readers interested in the general biology and biogeography of macroalgae are referred to several textbooks treating these subjects in depth (e.g. those of Dring 1982, Lobban and Harrison 1996, Dawes 1998, Graham, Graham and Wilcox 2009, and Wiencke and Bischof 2012).

2.4.1 *The green algae*

Those green algae (Division Chlorophyta) that are macroscopic have, like their microscopic counterparts, chlorophylls *a* and *b* as their main photosynthetic pigments; thus accounting for their green colour. This also implies that they absorb mostly the blue and red wavelengths of solar radiation (see Chapters 4 and 5) while the middle of the spectrum, containing green photons, is reflected or transmitted back to our eyes. The green macroalgae are common in all parts of the world's oceans (and are also very common in freshwaters), and they include the truly global genus *Ulva* (see Figure 2.4a), which has also become a popular 'test organism' for photosynthetic studies (see Chapter 7 and several chapters in Part III). Furthermore, this genus also contains species that form sulphated polysaccharides in their cell walls, and there is recently a surge towards exploiting the use of those compounds (Google the word 'Ulvans' for the latest updates in this field). It is conceivable that the green algae were the forefathers/mothers of terrestrial plants, which found their way out

Figure 2.4

Figure 2.4 (*Continued*) **Marine macroalgae.** The ubiquitous green algal genus *Ulva* (a) may take several forms of either broad, flat thalli (*Ulva rigida*, left) or tubular thalli (*Ulva flexuosa*, right, previously called *Enteromorpha flexuosa*). (b) The *Ulva* spp. have one large cup-shaped chloroplast in each cell. (c) *Codium geppiorum* is a coenocytic green alga; some *Codium* species (as well as another coenocytic genus, *Udotea*) are found in very deep waters. (d) Brown algae of the genus *Sargassum* are very common to most seas. Sometimes the larger forms of brown algae, e.g. *Laminaria saccharina*, form forest-like structures (e). Red algae can feature regular thallus shapes (like the *Cryptonemia* depicted in f) or can be encrusting, sometimes forming beds of rhodolites (g). Some red algae are farmed for the production of agar or carrageenan; the latter product is extracted from '*Eucheuma*' (h), here depicted as being reared in Zanzibar by a local woman (left) and then harvested and floated to shore by flat-bottomed boats (right). See further in the below text. Photos with permission from, and thanks to, Katrin Österlund (a, c–f and g), Ramon Bouchet-Roullard, Monterey Bay Aquarium Research Institute, California, USA (b), and Sven Beer (h).

of the oceans some 400 million years ago. Evidence for this relationship between green algae and higher plants includes the facts that a) both plant groups contain chlorophylls *a* and *b* as their main photosynthetic pigments, as well as the carotenoid β-carotene, and b) both groups are able to store surplus photosynthetic products as starch within their chloroplasts, sometimes to the extent that the entire plastid becomes white (the latter is then called an amyloplast; amylon = starch).

Many marine green macroalgae feature thin thalli in which most cells contain chloroplasts and are, therefore, photosynthetic. Some, e.g. *Ulva lactuca*, consist of two layers of photosynthetic cells (each containing one large, cup-shaped, chloroplast, see Figure 2.4b) while others form single-cell-layer tissue sheets (e.g. *Monostroma*) or tubes (e.g. *Ulva intestinalis*). Arguably, because of their thin thallus structure featuring an extremely high photosynthetic-surface to volume ratio, together with their photosynthetic traits (see Section 7.3), the opportunistically growing *Ulva* spp. (sp. and spp. stands for one or several, respectively, undefined species) feature the world's highest photosynthetic rates per plant area. Among the thicker-bodied green algae, some are coenocytic, i.e. consist of cells without separating cell walls or cell membranes. Some of these algae (e.g. *Codium* spp., see Figure 2.4c) can easily be used for chloroplast separations since their cellular contents can simply be squeezed out of a fractured thallus. Other green algae are calcareous, i.e. they form and deposit $CaCO_3$ (calcium carbonate) between their cells. Among them, the genus *Halimeda* is both coenocytic and calcareous, and has been used as a model plant for evaluating the process of calcification in relation to photosynthesis (see later sections).

2.4.2 The brown algae

Since all plants contain chlorophyll *a*, so do the brown algae (Division Phaeophyta). However, in addition to this photosynthetic pigment, they also contain chlorophyll *c* and the xanthophyll fucoxanthin (a type of carotenoid, see Section 4.2), which lends them their brownish colour. Fucoxanthin (like the chlorophylls) is located in the thylakoid membranes and functions in light capturing of the 450–540 nm wavelengths, peaking at 510–525 nm, thus complementing the chlorophylls in that region of the spectrum. Unlike the green algae, the browns are almost exclusively marine, and they include some of the largest plants on Earth: the giant kelps. These plants, belonging to the order *Laminariales* (lamina = thin plate or sheet, characteristic of these algae's blades; *Laminaria*, Figure 2.4e, is a common kelp genus in temperate seas), can reach lengths of about 50 m as in the genus *Macrocystis*; the latter forms forest-like underwater landscapes, e.g. off the Californian coast. The brown algae are not considered to be ancestors of terrestrial plants because their pigments and chloroplast structure differ from those of the higher plants. However, an interesting parallel evolution between the kelps and higher plants is found in that both contain vascular tissue (vasculum = small vessel) in their stems (in the kelps these are called stipes rather than stems). While the higher plants have both water- and photosynthate- (products formed by photosynthesis, mainly sucrose) conducting vessels in their vascular bundles (called xylem and phloem, respectively), the kelps have only phloem-like tissues. These tissues thus transport photosynthate (here mainly the reduced sugar mannitol) from the well-lit upper part of the kelp thalli to those lower parts (sometimes tens of metres down) that do not receive enough light to perform net positive diel photosynthesis. Another peculiarity of the brown algae is that some members of one genus, *Sargassum*, live free-floating in the Sargasso Sea (while other species of that genus are attached to rocky substrates just like the rest of the macroalgae). Thus, the free-floating forms can strictly speaking be seen as planktonic (illustrating that not all phytoplankton are of small size, just like not all zooplankton, e.g. up to metre-diameter jellyfish, are small).

As is the case for many of the red algae, some brown algae (notably from the *Fucales* and the kelps) are harvested and used directly for food or, more often for the brown algae, for the extraction of alginates. The alginates have a gelling capacity that makes them suitable for additions to food products that need thickening. Because of their high mineral content (especially of potassium, K), some brown algae common to North American, northern European and South African shores (notably *Fucus*, *Ascophyllum* and *Laminaria*, see Figure 2.4e) are used as natural fertilizers in agriculture. The abundance of those genera has also caused them to be appropriate subjects for much marine photosynthesis research (see Sections 7.3 and 10.2).

2.4.3 The red algae

The largely marine group of red algae (Division Rhodophyta) contains, in addition to chlorophylls *a* and *c*, other photosynthetic pigments collectively termed phycobilins (see Section 4.3). The fact that cyanobacteria also contain these pigments points towards the evolutionary relationship between those two groups: It is thought that red algae are early descendants of cyanobacteria (following endosymbiosis, see Box 1.3) within the 'red lineage' of plant evolution in the oceans (see the book Aquatic Photosynthesis by Paul Falkowski and John Raven for a detailed account of the evolution of marine photosynthetic organisms).

As in the case with the cyanobacteria, the red algae contain two types of phycobilins: the red-coloured phycoerythrin and the bluish phycocyanin. However, while phycocyanin is the main pigment of most cyanobacteria, phycoerythrin is present in relatively higher amounts in the red algae. These pigments are located on the thylakoid membranes in small structures called phycobilisomes (cf. Figure 4.3), which thus mask the green colour of chlorophyll embedded within the membranes. However, while phycoerythrin usually colours the red algae red, the green colour of chlorophyll sometimes competes with it in appearance; if phycoerythrin is present at low concentrations, the thallus of a red alga may look brownish (the left-hand picture of Figure 2.4h) or greenish (the lower levels of algae in the right-hand picture of Figure 2.4h), and only the extraction and verification of the presence of phycobilins will confirm that those algae actually belongs to the Rhodophyta. While phycoerythrin is an accessory pigment to chlorophyll in light capture for photosynthesis, this role seems to be especially important under low-light conditions. Therefore, the concentration of phycobilins is often low under high irradiances. Thus, many intertidal red algae look green in their exterior parts that are exposed to full sunlight but become successively redder towards their holdfast to where less light penetrates.

While some green algae (and one brown algal genus, *Padina*) are calcareous, the plant-world of calcifiers belongs to the red algae. These are within the group Coralllinaceae, sometimes called the coralline algae because, like corals, they deposit $CaCO_3$ (in the form of either aragonite or calcite, see Box 2.3, though corals deposit aragonite only). These algae may coat the rocks on which they grow (in the case of encrusting species), they may cement together parts of coral reefs, or they may even deposit most of the $CaCO_3$ after they die, thus forming entire coralline-algal-based reef-like structures (largely in analogy with the deposition of $CaCO_3$ by reef-forming corals). In other cases extensive seafloor areas are covered with free-living coralline algae called rhodoliths or 'maerl' (see Figure 2.4g) that can be several metres thick. See also Box 2.3 regarding calcifying algae; the relationship between photosynthesis and calcification will be treated in Section 7.5.

One particular feature of some red algae is that their cell walls contain the sulphated polysaccharides agar or carrageenan, the building blocks of which are galactose (instead of the glucose found in cellulose). These algae are increasingly being farmed in maricultural ventures; in particular the agarophyte *Gracilaria* and the carrageenophyte *Eucheuma* (see Figure 2.4h) are grown for those polysaccharides, and their photosynthetic traits are also being investigated in order to optimise growth rates (some relationships, and non-relationships, between algal photosynthesis and growth will be discussed briefly in Section 12.3). While these algal products are used in both the

Box 2.3   Calcification in algae

Many marine algae, both macro- and micro-scopic forms, deposit crystals of calcium carbonate ($CaCO_3$) in their cell walls. This gives them a hard structure that probably makes them less attractive for grazers. The $CaCO_3$ left behind after they die may be an important component in structuring reefs, or they may deposit it as sand grains, but more about the latter in Section 11.3. Regarding microalgae, calcification by the coccolithophores is hugely important in the global carbon cycle, producing ~12% of the global oceanic $CaCO_3$ accumulation of 1.1 Gt y^{-1}. Some dinoflagellates also produce calcium carbonate, but are of lesser importance ($\sim$ 3.5%) in terms of global biogeo-chemistry of carbon. Coccolithophores produce calcium carbonate (mostly in the form of calcite, see below) scales/plates in vesicles inside the cell, and these plates (coccoliths) are then extruded to the outside of the cell (see Figure 2.5a). Coccolithophores can form huge surface blooms in the oceans with anything up to 1 tonne of calcite km^{-2} in such blooms.

All macroalgal groups contain genera that deposit $CaCO_3$. However, only one genus of brown alga, *Padina*, does so. Amongst the green algae, the major calcifying genus is *Halimeda* (see Figure 2.5c) commonly found in most tropical and subtropical waters. Species of this genus are built up as jointed hard, flattened, leaf-like thallus segments that normally contain 60–80% $CaCO_3$ as aragonite[7]. Decomposed *Halimeda* tissues are important sediment elements in the areas where they are present, and the major part of the calcareous sediment found in the sandy beaches of many tropical shores, can be made up by disintegrated *Halimeda* parts (see Section 11.3). Other important sediment producers within the green algae are *Udotea* and *Penicillus*.

Within the red algae, most calcifying species are coralline, i.e. marine red macroalgae of the family Corallinaceae, all of which deposit $CaCO_3$ as calcite[7]. Crustose coralline algae build cementing layers covering and protecting coral reefs and offer substrates for coral settlements. Many calcareous algae also grow epiphytically on seagrass leaves as well as on other marine organisms such as bivalves and molluscs. See more about the mechanisms of algal calcification in Section 7.5.

Why do some calcifying algae form calcite while others form aragonite? The full reason is not known, but has probably something to do with temperature, water chemistry and the organic matrix in or around which the $CaCO_3$ crystals are formed. In this respect, the ratio between Mg^{2+} and Ca^{2+} has a strong influence, and the relatively high Mg^{2+} content of seawater (~30 mM) seems to favour aragonite. However, as said, the red coralline algae deposit calcite within their cell wall, and it is possible that the nature of their cell wall governs the formation of this isomorph[7].

---

[7]Calcium carbonate ($CaCO_3$) can appear in two different forms: aragonite and calcite. Both have the same chemical composition (they are 'isomorphs'), but with different crystalline structures. In aragonite the molecules are arranged in an 'orthorhombic' symmetry, whereas calcite has a 'trigonal' symmetry, making the latter form more stable. Aragonite is denser, but still more soluble than calcite. The calcite formed by algae can also contain substantial amounts of magnesium in the form of $MgCO_3$, or 'magnesian calcite', making it more soluble than pure $CaCO_3$.

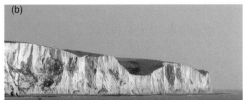

Figure 2.5   Calcifying algae. A scanning electron micrograph of the coccolithophore *Emiliania huxleyi* featuring calcium carbonate ($CaCO_3$) scales called coccoliths (a) (this strain was originally isolated from Pipe Lagoon in Tasmania, Australia). Note the sculpted coccoliths on the surface; these are produced internally in vesicles and extruded to the surface (see Section 7.5). The coccoliths of this organism have, during millions of years, deposited thick sediment layers such as found today, e.g. as the famous Cliffs of Dover (b). The calcifying green alga *Halimeda* (c) shows here various stages of thallus deterioration, from the live and healthy plant (left), through dead and partially disintegrated thalli (middle) and the remaining $CaCO_3$ that constitutes the gravel and sand in some tropical bays (see Section 11.4). Some coralline red algae form the rhodoliths depicted in (d). Photos by Slobodanka Stojkovic (a), Sven Beer (c) and Mats Björk (d). Picture (b) was taken from Wikipedia (http://upload.wikimedia.org/wikipedia/commons/7/7e/White_cliffs_of_dover_09_2004.jpg).

pharmaceutical and food industries as gelling agents (e.g. almost every biology student will have grown bacteria on agar-containing Petri dishes), some red algae are also eaten more or less as they are. The classical example of edible algae is *Porphyra* (called Nori in Japan), which is used to wrap around the rice in sushi. (Incidentally, *Porphyra* differs from most red algae by growing in temperate regions such as Japan; most other red algae are tropical.) This alga has also been the subject of considerable photosynthetic investigations (see, e.g. Section 10.1).

2.5   Seagrasses

The macroalgae described above can be considered as macrophytes (again, 'phytos' = plant, and we consider all algae as functionally being plants), but there is also a starkly different group of marine macrophytes that have attracted relatively less attention. This is a group of angiosperms, or flowering plants, which found their way back to the oceans long after the plants had invaded land some 350 million years ago. Thus, 90–100 million years ago, when there was a rise in seawater levels, some of the grasses that grew close to the seashores found themselves submerged in seawater. One piece of evidence that supports their terrestrial origin can be seen in the fact that residues of stomata can be found at the base of the leaves. In terrestrial plants, the stomata restrict water loss from the leaves, but since seagrasses are principally submerged in a liquid medium, the stomata became absent in the bulk parts of the leaves. These marine angiosperms, or seagrasses, thus evolved from those coastal grasses that successfully managed to adapt to being submerged in saline waters. Another theory has it that the ancestors of seagrasses were freshwater plants that, therefore, only had to adapt to water of a higher salinity. In both cases, the seagrasses exemplify a successful re-adaptation to marine life (in analogy with, e.g. whales and sea-turtles).

While there may exist some 20 000 or more species of macroalgae (not all have yet been identified, but see also Box 2.4 on taxonomical uncertainties in marine plants), there are only some 50 species of seagrasses, most of which are found in tropical seas. (See Figure 2.6 for some common forms.) These 50 species belong to 10 genera, all in turn belonging to the monocotyledonous higher plants. Like their terrestrial counterparts, the seagrasses feature structures such as roots (and, like most grasses, also rhizomes[8]), leaves and flowers, and can undergo pollination and produce seeds (though most species rely heavily on vegetative spread). Seagrasses may grow in large meadows (Figure I.1e) that, because of their vegetative growth, can be comprised of one (or a few) single clone(s). Our knowledge of seagrasses was furthered considerably by Cees den Hartog in his book Seagrasses of the World.

The roots of seagrasses, like those of terrestrial higher plants, function in nutrient uptake. The leaves can also take up nutrients from the water column, but the ability to extract nutrients from the sediment renders the seagrasses at an advantage over (the root-less) macroalgae in nutrient-poor waters. (For example, the nutrient-poor, clear, waters of the Red Sea probably contain a higher biomass of seagrasses than of macroalgae, while the opposite is true for the nutrient-rich, murky, waters of, e.g. the Baltic Sea). Unlike the often nutrient-poor waters where seagrasses thrive, the sediments that support seagrass growth are often

[8]Rhizomes are horizontal stems within the sediments that connect the various individual shoots (usually comprised of bundles of leaves). Seagrasses can multiply sexually by pollination and setting of seeds, but often spread clonally by rhizome elongation and, so, an entire sea grass bed of up to tens of m^2 or more can be one genetic clone in which rhizomes bud off new shoots.

Figure 2.6   Seagrasses. The Indo-Pacific seagrass *Halophila stipulacea* (a). A leaf cross section of *Cymodocea rotundata* showing lacunae (b); also seen are chloroplasts, which are mainly located in the epidermal cells. *Cymodocea serrulata* as seen with some uncovered roots and rhizomes (c), exemplifying their role in stabilizing the sediment and preventing erosion. The temperate species *Zostera marina*, here shown as partly covered by filamentous algae (d), which is a common phenomenon in eutrophic waters. Details of *Cymodocea nodosa* shoots, showing the uncovered rhizome and roots (e). Photos by Mats Björk (a, c–e) and Fred Short (b).

rich in nutrients. It follows from that, however, that the sediments are also poor in O_2 due to the high microbial activity therein. In order for roots to grow in such an O_2-depleted medium, the seagrasses have evolved internal aeration canals called lacunae that supply the roots with O_2 (see Figure 2.6b). These lacunae are thus air canals through which part of the O_2 produced photosynthetically during the day is conducted from the leaves towards the roots (while another part of the O_2 diffuses out into the water column). This transport system is so efficient that even zones of the sediment surrounding the root tips become oxidised. While the lacunae form a continuous, vertically oriented, gas-transport system throughout seagrass plants, they also smartly contain septa through which liquids cannot pass. In this way, if a seagrass leaf is chopped off by, e.g. turtle or fish grazing, seawater will not flood the lacunar system, which would suffocate the roots and, consequently, kill the whole shoot. Instead, water enters only a very small fraction of the leaf, and the underlying tissue can continue to function and grow (the meristems from which the leaves grow are located at the base of the leaves).

Another important role of seagrass roots is their anchoring of the plants to the sediment. Thus, the roots also stabilise the sediment (see Figure 2.6c), and erosion has been shown to occur in areas where formerly lush seagrass growth has been decimated. Indeed, one of the basic differences in habitat utilisation between macroalgae and seagrasses is that the former usually grow on rocky substrates where they are held in place by their holdfasts, while seagrasses inhabit softer sediments where they are held in place by their root systems.

Unlike macroalgae, where the whole plant surface is photosynthetically active, large proportions of seagrass plants are comprised of the non-photosynthetic roots and rhizomes. Therefore, the leaves of seagrasses also have to support those tissues with photosynthate (= products of photosynthesis, mainly sugars). This means, e.g. that seagrasses need more light in order to survive than do many algae (see further about light utilisation strategies of marine plants in Chapter 9).

Seagrasses are closely related to terrestrial grasses taxonomically, but are very different from them ecologically. Partly therefore, photosynthetic research has focused on their carbon acquisition in the marine aquatic medium (see Section 7.4). Furthermore, since some seagrasses can grow intertidally, and thus spend part of their time exposed to air, and given their lack of stomata, other studies have focused on elucidating their photosynthetic behaviour under the threat of desiccation (Section 10.1). Thirdly, recent evidence suggests that seagrasses, through their photosynthesis, can dictate conditions in their surrounding water that can affect other organisms, and this, together with their potential role in making entire ecosystems resilient to, e.g. ocean acidification, will be treated in Section 11.4.

Box 2.4   The rise and, sometimes, fall in numbers of species

In general, the number of recognised genera and species on our planet increases as new ones are discovered. This is quite natural since there is no doubt that new life forms will be found as new regions of, e.g. the deep oceans are explored. However, sometimes taxonomists seem too eager to describe forms of existing taxa not earlier illustrated as new ones. Two examples can be given from the marine-plant realm: In the first, the seagrass *Halophila ovalis* can feature various leaf forms, often related to the environment in which it grows. Based on these differences, the number of *Halophila* species described has increased during the last 20 years; the genus became the most species-rich among seagrasses, and the total number of seagrass species increased to well over 50. However, molecular techniques in which DNA sequences are compared, have often shown such strong similarities between the different forms of *Halophila* that their differentiation into different species is unwarranted. Thus, the number of seagrass species has fluctuated, but is, interestingly, following a decreasing trend. Another example is from the green algae of the family Ulvales. Until some 10 years ago, the genus *Ulva* was held as being distinct from the genus *Enteromorpha*; the former is characterised by a double layer of photosynthesising cells while the latter is tube-shaped with a single cell-layer. A DNA analysis of the two forms shows, however, that there is no valid ground for distinguishing between the two genera, and therefore the genus *Enteromorpha* is now history as it has been absorbed into the genus *Ulva* (it was also originally described by Linnaeus as such). For example, the formerly common species *Enteromorpha intestinalis* is now called *Ulva intestinalis*. These two cases exemplify that the number of taxa within a certain group of organisms is not always expanded, but can also become reduced (for the great benefit to those, read 'us', i.e. the authors, that have a hard time remembering names, and Latin names in particular).

Chapter 3

Seawater as a medium for photosynthesis and plant growth

The oceans provide, of course, very different surroundings for plant growth than do terrestrial environments. However, it must be remembered that this is true only for the external medium; all cellular processes such as photosynthesis take place in the aquatic micro-environments of the cells, and a cell can therefore be seen as a micro-ocean (albeit of a lower salinity than most seas). External to the marine plant, water, because of its much higher density than air, physically supports the plants growing therein. This is true for the (usually) microscopic phytoplankton (see the previous chapter), which can drift around in the upper water layers without sinking to the bottom while alive, as well as for the much larger macrophytes (i.e. macroalgae and seagrasses, also see the previous chapter), the (usually) anchored bodies of which are largely supported by their surrounding water such that they do not 'need' to invest in producing as much structural material as terrestrial plants (the latter of which

need, e.g. stems to hold their photosynthesising leaves up from the ground and facing the sun). Indeed, marine plants usually contain less structural tissues than their terrestrial counterparts – this can easily be seen when picking up a macroalga, which will often become floppy and crumples down when held up in the air.

One main difference between terrestrial and aquatic environments pertaining to biochemical processes such as photosynthesis is that the **rate of diffusion** is some 10 000 times slower in water than in air. This means that, e.g. the diffusional supply of CO_2 to the plant surface is restricted in water. Diffusional supply means that even if the water surrounding a marine plant is agitated and the supply of CO_2 towards the plant by bulk flow is relatively fast, there is always a water layer around the plant that is unstirred and through which this nutrient must pass by diffusion. (In contrast, the diffusion through an unstirred layer close to a terrestrial plant is, as noted, some

10 000 times faster.) Similarly, the efflux of O_2 generated by photosynthesis is hindered by that same liquid unstirred layer surrounding a marine plant (and we will see in Chapter 6 that excessive O_2 accumulation within photosynthesising cells may inhibit photosynthesis). The **unstirred layer** through which supply to, and export from, the cells is only via diffusion is, accordingly, called the **diffusion boundary layer** (DBL), and it will be discussed in relation to photosynthetic mechanisms in marine plants in Chapter 7. Other environmental conditions that are relevant to the successful growth and photosynthesis of marine plants will be detailed in the following text, and special emphasis will be given to light and CO_2 conditions, which differ greatly between marine and terrestrial environments.

3.1 Light

While light is the major driving force of photosynthesis, the definition of 'light' is not self-evident (see SB's Personal Note below). However, if we define 'visible light' as the elec-tromagnetic wave upon which those energy-containing particles called **quanta** 'ride' that cause vision in higher animals (those quanta are also called **photons**) and compare it with light that causes photosynthesis, we find, interestingly, that the two processes use approximately the same wavelengths: While mammals largely use the 380–750 nm (nm = 10^{-9} m) wavelength band for vision, plants use the 400–700-nm band for photosynthesis; the latter is therefore also termed **photosynthetically active radiation** (**PAR**; see Figure 3.1).

One property of light that is important for the further discussion of both its attenuation in seawater (below) and its utilisation in photosynthesis (Chapter 5) is that the energy (E) of a photon is inversely proportional to its wavelength (λ; see also Box 3.1). This can be seen in the general relationship $E = hc/\lambda$, where h is Planck's constant (6.63×10^{-34} J·s), c is the speed of light ($\sim 3 \times 10^8$ m s^{-1}) and λ (the Greek letter lambda) is the wavelength. If λ is given in μm (10^{-6} m) and h in electron volts (eV, i.e. the energy needed to move an electron through 1 V) instead of J·s, then a practical way of calculating the energy of a photon is $E(eV) = 1.24/\lambda$.

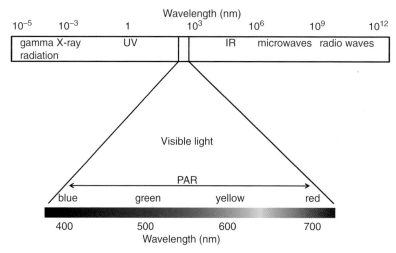

Figure 3.1 **The electromagnetic spectrum** (upper horizontal bar), **visible light** (lower horizontal bar) **and photosynthetically active radiation** (PAR, double-sided arrow). Note that 10^9 nm = metre; radio waves can thus be up to kilometres in length. Drawing by Sven Beer.

Personal Note: What is light?

At the appropriate time I (SB) ask the students of my undergraduate Marine Botany course "So what is light? Who can define 'light'?" Silence! "Come on, it's not as if I'm asking you to define 'time'." After some thought, one student says "light is energy." "Yes, but there are many different kinds of energy..." "OK, light is an electromagnetic wave ... but it's also particles, quanta I think they called them in physics class." I show them a slide of the entire electromagnetic spectrum (like in the upper part of Figure 3.1): "But your definition encompasses all electromagnetic radiation, including X-rays, microwaves and even radio waves," and then I let PowerPoint 'box out' the 380–750 nm band of the electromagnetic spectrum (like in the lower part of Figure 3.1). "Ah, light is the 380 to 750 nm portion of the electromagnetic spectrum" says another student. "Yes, but why is this portion of the spectrum called 'light'?" Silence. "OK, I'll tell you: It is called light because this is the part of the spectrum that can be perceived by the pigments in the retina of our eyes, and it triggers a flow of information to our brain that can be perceived as the picture we see." So, we decide on 'light' being that part of the electromagnetic spectrum on which those quanta 'ride' that contain the appropriate energy to cause vision in animals. We then go on to define the light that plants 'see' (see the text pertaining to photosynthetically active radiation, PAR). If a student asks "but how come that animals and plants use almost identical wavelengths of radiation for so very different purposes?", my answer is "sorry, but we don't have the time to discuss that now", meaning that while I think it has to do with too high and too low quantum energies below and above those wavelengths, I really don't know.

—Sven Beer

However, λ is often for practical reasons given in nm (10^{-9} m, e.g. as in Figure 3.1), and since the absolute energy of a photon means rather little to the average biologist (such as ourselves), it suffices for the level of this book to realise that a blue photon of 400 nm wavelength contains almost double the energy of a red one of 700 nm, while the photons of PAR between those two extremes carry decreasing energies as wavelengths increase. Accordingly, low-energy photons (i.e. of high wavelengths, e.g. those of reddish light) are absorbed to a greater extent by water molecules along a depth gradient than are photons of higher energy (i.e. lower wavelengths, e.g. bluish light), and so the latter penetrate deeper down in clear oceanic waters (see below). In the first steps of light utilisation for photosynthesis, however, the different energies of photons are 'adjusted' to a certain level before being used for the process itself (see Section 5.1). Consequently, it is the **number of photons** absorbed by the photosynthetic pigments rather than their energy content that is of importance for the overall process of photosynthesis. For this reason, the irradiance unit of choice when referring to plants' utilisation of light is μ**mol m^{-2} s^{-1}** (a mol being, as always, 6.023×10^{23} 'units', or in this case photons, and a μmol being, accordingly, 6.023×10^{17} 'units', or in this case photons; the surface irradiance on a sunny day is ~2000 μ**mol photons m^{-2} s^{-1}** PAR)[1].

---

[1]One mol of photons was previously called one Einstein (E), and a μmol of photons a microEinstein (μE). However, a) those are not acceptable SI units and b) the greatness of Albert Einstein is not compatible with the prefix 'micro', and so the Einstein units are no longer used for irradiance or photon flux (both terms will be used synonymously in this book, but we use 'irradiance' more).

Box 3.1   A(nother) perception of light

Light can be seen as electromagnetic waves of those wavelengths that stimulate the pigments in our retinas so as to produce sight (see the text for this and note the slight difference between light for sight and irradiance for photosynthesis). These waves that move ahead at the speed of light are said to have two perpendicular dimensions: one of electricity and one of magnetism (together forming the 'electromagnetic' wave). The simple relationship between the speed (c), frequency (v) and wavelength (λ) of light is $c = v \cdot \lambda$, i.e. the wavelength is inversely proportional to the frequency since the speed of light is constant. We think that the wavelength, λ, is easier to comprehend than the frequency, and will therefore use it extensively in this text for describing the quality of light. The energy (E) of a photon is proportional to its frequency and, then by definition, inversely proportional to its wavelength (as described in the above text). So, for example, a photon of red light (~700 nm) carries a little more than half the energy of a blue photon (~400 nm).

In water, the spectral distribution of PAR reaching a plant is different from that on land. This is because water not only attenuates the light intensity (or, more correctly, the photon flux, or **irradiance**, a term that we will stick with much in this book), but, as mentioned above and detailed below, the attenuation with depth is wavelength dependent; therefore, plants living in the oceans will receive different spectra of light dependent on depth (as well as other physical properties of the water, see below). The two main characteristics of seawater that determine the quantity and quality of the irradiance penetrating to a certain depth are absorption and scatter.

Light **absorption** in the oceans is a property of the water molecules, which absorb photons according to their energy (the latter of which is converted to heat). Thus, red photons of low energy are more readily absorbed than, e.g. blue ones; only <1% of the incident red photons (calculated for 650 nm) penetrate to 20 m depth in clear waters while some 60% of the blue photons (450 nm) remain at that depth. As a result, so few red photons remain at (or below) 20 m depth that red colour cannot be perceived there by, e.g. a diver (unless he/she takes some red photons along

as part of the white light from a flashlight), and the surrounding looks bluish. The same goes for underwater photography in which flash-less pictures from deep waters always turn out bluish (unless corrected by picture-editing programs such as Photoshop©). So, in summary, in clear waters where the spectrum is dependent mainly on absorption, the surroundings turn increasingly bluish with depth. In some cases, not only the water molecules but also coloured substances may absorb certain wavelengths selectively. For example, coastal waters may appear brownish-yellow because of the presence of dissolved substances that result from decomposing vegetation both in the sea and on nearby lands (from where they are washed into the sea); these substances are sometimes called 'Gelbstoff' (from the German gelb = yellow; stoff = 'stuff'); see, e.g. Figure 3.2.

Scatter, the other property that attenuates light in water, is mainly caused by particles suspended in the water column (rather than by the water molecules themselves, although they too scatter light a little). Unlike absorption, scatter affects short-wavelength photons more than long-wavelength ones (see Figure 3.2). (If scatter is envisioned as photons 'bouncing off'

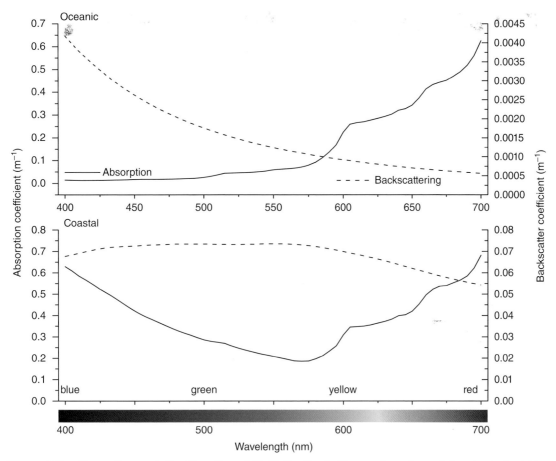

Figure 3.2   Absorption and (back)scatter of light for 'typical' oceanic and coastal waters, the latter being influenced by organic substances from land runoffs (see the high absorption towards blue). Note that although more than an order of magnitude lower, there is still some scatter of blue light in oceanic waters, enabling the oceans to appear blue (see Box 3.2). Courtesy of Charles Gallegos.

particles, then one can further envision that shorter wavelengths have a higher chance of 'bumping into' a tiny particle than have longer wavelengths, the latter of which may 'squiggle around' it.) Thus, in turbid waters, photons of decreasing wavelengths are increasingly scattered. Since water molecules are naturally also present, they absorb the higher wavelengths, and the colours penetrating deepest in turbid waters are those between the highly scattered blue and highly absorbed red, e.g. green. The greenish colour of many coastal waters is therefore often due not only to the presence

of chlorophyll-containing phytoplankton, but because, again, reddish photons are absorbed, bluish photons are scattered, and the mid-spectrum (i.e. green) fills the bulk part of the water column.

The zone that extends vertically from the sea surface down to the depth where enough light remains so as to support plant growth is called the **euphotic zone** (eu- = well; photic = lit). Classically, the irradiance level at the deep end of the euphotic zone has been set to (based on early observations) **1%** of the surface irradiance (see Figure 3.3), but lately many

Box 3.2   Why are the oceans blue?

People, including marine-science students, often ask why the sea is blue. First, the sea is not always blue: as mentioned in the text, many coastal waters appear greenish or brownish-yellow. However, the open ocean, several kilometres or miles from the shore, almost always appears as blue. The reason for this is that in unpolluted, particle-free, waters, the preferential absorption of long-wavelength (low-energy) photons is what mainly determines the spectral distribution of light attenuation. Thus, short-wavelength (high-energy) bluish photons penetrate deepest and 'fill up' the bulk of the water column with their colour. Since water molecules also scatter a small proportion of those photons (but much less so than other particles in, e.g. coastal waters, see Figure 3.2), it follows that these largely water-penetrating photons are eventually also reflected back to our eyes. Or, in other words, out of the very low scattering in clear oceanic waters, the photons available to be scattered and, thus, reflected to our eyes, are mainly the bluish ones, and that is why the clear deep oceans look blue. (It is often said that the oceans are blue because the blue sky is reflected by the water surface. However, sailors will testify to the truism that the oceans are also deep blue in heavily overcast weathers, and so that explanation of the general blueness of the oceans is not valid.)

exceptions have been found. For example, sea-grasses need much more light for their survival than 1% of surface light. This is because they need enough light to produce enough photosynthate to support not only their green parts, but also their extensive non-green root and rhizome tissues that grow in the sediment. On the other hand, many small cyanobacteria can get by with much less than 1% of the surface irradiance; some of the very tiny, nanoplanktonic, types may need as little as 0.1% or less! (See more on light requirements in Chapter 9 of this book.)

One parameter instrumental for characterising the amount (irradiance) and quality (spectrum) of light underwater is the **attenuation coefficient**, K_d. The K_d can further be divided into the absorption coefficient and back-scatter coefficient as in Figure 3.2. While the absorption by water remains inversely proportional to the wavelength, the various coloured substances will add to the absorption especially in coastal waters such as seen in the case depicted in Figure 3.2. Here, the water contains chlorophyll, dissolved organic matter ('Gelbstoff') and such particulate matter that can also cause absorption of light, especially of short wavelengths. In such waters, the backscatter is much more evenly distributed along the wavelengths of PAR than in open, oceanic, waters. Since it requires expertise and sophisticated equipment to determine wavelength-specific absorption and scatter in water bodies, we will here restrict the text to include the simple determination of an overall K_d for PAR in a given water body by measuring PAR irradiance at two depths and then using a formula to calculate K_d (see Box 3.3); knowing this K_d value and the incident surface irradiance, the irradiance at any depth along that water body can then be calculated from the Beer- (no relation to Sven)Lambert law. For more information on light penetration into oceanic waters, we refer to the textbook Light & Photosynthesis in Aquatic Ecosystems by John Kirk.

When discussing light requirements for plant growth, it must be remembered that they should strictly not be those that cause

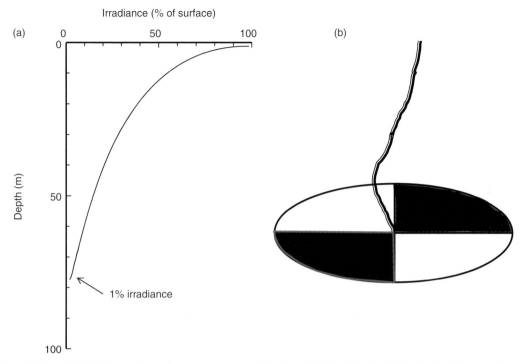

Figure 3.3   (a) **Light (irradiance) as a function of depth**, with the 1%-limit (the limit of the photic zone for many phytoplankton and macroalgae) marked (assuming an absorption coefficient, K_d, of ~0.05). Note the exponential attenuation of light with depth. (b) A **Secchi disc** (with a line for lowering it down) with which the water clarity can be estimated (see Box 3.3). Drawing by Sven Beer.

photosynthesis at a certain time of the day (e.g. the irradiance during midday) but, rather, the daily accumulated irradiance at which photosynthesis during the entire day outweighs respiration during both day and night. Thus, it is the diel gas-exchange balance of net CO_2 uptake (i.e. the net result of CO_2-fixation and -reduction and respiration) or O_2 evolution (resulting from photosynthesis and respiration combined) that must be positive in order for the plant to survive and grow (see more about net, or apparent, and gross, or true, photosynthetic gas exchange and diel cycles of photosynthesis in Chapter 9). Having said that, yes, many light requirements of specific plants have been calibrated to, e.g. the maximum daily (e.g. during midday) irradiance at a certain depth, or to how many hours of saturating light (for

photosynthesis) a plant needs in order to grow at a certain site.

3.2   Inorganic carbon

Carbon dioxide (CO_2) in the air (i.e. atmospheric CO_2) is the only external form of inorganic carbon (Ci) available for photosynthesis in terrestrial plants. However, in marine environments CO_2 not only dissolves in seawater, but also reacts with the water molecules so as to form carbonic acid (H_2CO_3), which dissociates to bicarbonate (HCO_3^-) and carbonate (CO_3^{2-}) ions and protons (H^+). As we shall see in Chapter 7, HCO_3^- is the main Ci form from the bulk medium that is used for photosynthesis in most marine plants. (It

Box 3.3   Calculating irradiance at depth, and use of the Secchi disc

The attenuation coefficient, K_d, for PAR in general can easily be determined for a certain water type. All that is needed is a light meter measuring PAR irradiance (400–700 nm), preferably as photon flux (i.e. in µmol photons m^{-2} s^{-1}), and the measurements can be done as follows: I_{Z1}, the irradiance at depth Z_1, and I_{Z2} at a deeper depth, Z_2, is measured at whichever depths are convenient (e.g. Z_1 can be 0.5 m and Z_2 1 m). The attenuation coefficient can then be calculated as:

$$K_d = 1/(Z_2 - Z_1) \cdot \ln(I_{Z2}/I_{Z1}) \qquad (3.1)$$

After the K_d has been determined, the irradiance at any depth (I_x) can be calculated by the Beer–Lambert law as:

$$I_x = I_0 \cdot e^{-K_d \cdot x} \qquad (3.2)$$

where I_0 is the surface irradiance preferable measured just below, but as close as possible to, the water surface. For example, in the above example, if Z_1 and Z_2 equal 1000 and 300 µmol photons m^{-2} s^{-1} at 0.5 and 1 m depth, respectively, then equation (3.1) will yield a K_d value of 0.6. Inserting this number into equation (3.2) for calculating, e.g. the irradiance at 10 m depth, and given that the surface irradiance is 2,000 µmol photons m^{-2} s^{-1} (a sunny day!) results is 5 µmol photons m^{-2} s^{-1}, which is 0.25% of full sunlight. Thus, in this example, and assuming, e.g. that a macroalga needs 1% of full sunlight during midday in order to photosynthesise enough to show positive diel photosynthesis and growth, we arrive at the conclusion that the water at 10 m is too murky to sustain growth of that macroalga.

Today, irradiance under the water is measured by quantum sensors set to measure PAR (400–700 nm). However, before such instrumentation became commonly available, the more subjective method of using a so-called Sechi disk was popular. The Secchi disc was invented some 150 years ago by the Italian Pietro Secchi. Today's disks are about 20 cm in diameter and are painted with two white and two black quadrants (see Figure 3.3b). The disk is lowered down into the water column until one cannot see it any longer – this depth is called the Secchi depth. While Secchi disk readings for a certain water type do depend on the eyesight of the observer, there is often a remarkably good correlation between Secchi depths readings and irradiance measurements using, e.g. a PAR probe. In such cases, 1.44 divided by the Secchi depth will approximately yield the light attenuation coefficient, K_d (according to Kirk's book Light & Photosynthesis in Aquatic Ecosystems).

should be noted, though, that it is only CO_2 that is reduced in the photosynthetic reactions of the Calvin cycle, see Chapter 6.) While Ci, like the other factors discussed in this chapter, is abiotic (= not directly relating to biological origin), the biota surrounding a specific marine plant can alter the composition of this (as well as other) abiotic factor on a diel scale (see further in Part III of this book). On a longer time scale, 'global change' involves a steep increase in atmospheric CO_2 concentrations as observed for the last 100 years or so. As a consequence, the seawater-dissolved CO_2 concentration also increases proportionally, and the pH decreases, a phenomenon known as ocean acidification, and these changes, and their possible influence on marine plant photosynthesis and growth, will be discussed in Section 12.2.

At any given temperature and salinity, the equilibrium concentrations of dissolved and atmospheric CO_2 are directly proportional to one another (according to Henry's law). Carbon dioxide dissolves much more readily in water than does O_2 (see Box 1.1 above); for example, at 20 °C and a salinity of 35 (see below for salinity units), today's atmospheric CO_2 concentration of 380 ppm (or 380 μATM – oops, it just increased to 394 ppm, or ~18 μM[2], on this 13th of July, 2013) results in a dissolved concentration of 12.7 μM (see Reaction 3.4). At lower temperatures, the atmospheric and dissolved CO_2 concentrations are even closer to one another. Similarly, at higher salinities there is less CO_2 dissolved in the water at any given atmospheric CO_2 concentration (although there is more Ci altogether).

As mentioned above, CO_2 not only dissolves in seawater but also reacts with water to form H_2CO_3, which then dissociates to form HCO_3^- and CO_3^{2-} according to:

$$CO_{2(atm)} \leftrightarrow CO_{2(diss)} + H_2O \leftrightarrow H_2CO_3$$
$$\leftrightarrow HCO_3^- + H^+ \leftrightarrow CO_3^{2-} + 2H^+ \quad (3.3)$$

where 'atm' and 'diss' stand for atmospheric and dissolved CO_2, respectively. As indicated by the protons (H^+) in the reaction, the concentrations of the ionic Ci forms (HCO_3^- and CO_3^{2-}) in equilibrium with the other Ci forms depend on pH. In open waters of pH 8.1 (which is commonly found in today's oceans), the con-

centrations of the various Ci forms (in μM) in equilibrium with today's (July 2013) atmosphere containing 394 ppm CO_2 are as follows (at 20 °C and a salinity of 35, see below for salinity units):

$$\underset{\underset{17.5}{\mu M}}{CO_{2(atm)}} \leftrightarrow \underset{12.7}{CO_{2(diss)}} + H_2O \leftrightarrow \underset{<1}{H_2CO_3}$$
$$\leftrightarrow \underset{1547}{HCO_3^-} + H^+ \leftrightarrow \underset{132}{CO_3^{2-}} + 2H^+ \quad (3.4)$$

As can be seen, the equilibrium concentrations of the ionic Ci forms are, at the 'normal' seawater pH, much higher than that of the dissolved CO_2; the ca. 120 times higher concentration of HCO_3^- than of CO_2 is in fact the rationale for the former being the principal external Ci form used by most marine plants, but much more about that in Chapter 7. Also, it can be mentioned here (and will be elaborated on later in Chapter 7) that the $CO_{2(diss)} + H_2O \leftrightarrow H_2CO_3$ part of the overall reaction is slow, with a half-time (the time it would take for half an introduced amount of CO_2 to equilibrate with HCO_3^- or *vice versa*) of up to a minute. This, and why (and how) some marine plants must (and can) overcome such a potential limitation will also be further discussed in Chapter 7.

The relationship between the different Ci forms as a function of pH is often shown in graphs such as those depicted in Figure 3.4. This graph shows the **relative proportions** of the different Ci forms when the pH is changed by adding acids or bases in artificial systems or when pH is changing with additions or depletions of CO_2 by photosynthetic or respiratory activities. This figure can, however, not easily be used for depicting actual concentrations as a function of pH in natural systems because it does not take into consideration the changes in total Ci ($CO_2 + HCO_3^- + CO_3^{2-}$ + a little H_2CO_3) due to photosynthetic activities and respiration. If Figure 3.4 is to be used for open-system calculations of actual Ci concentrations,

[2]The relationship between volume- (e.g. ppm, parts per million) and molarity-based (e.g. μM) units is given by the ideal gas law: at 0 °C and 1 atmosphere pressure 1 mol of an 'ideal' gas occupies 22.4 L; at 20 °C it occupies ~24 L. Thus, the micromolarity (μM) of a gas can be approximated as its concentration in ppm divided by 24 (at 20 °C).

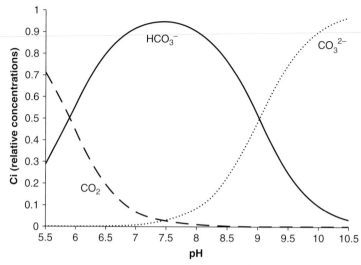

Figure 3.4 **The distribution of inorganic carbon** (Ci) **forms as a function of pH** (at 20 °C and a salinity of 35). Observe that the units are relative to 1.0, such that total Ci concentrations need to be known in order to calculate the actual concentrations of each Ci form. CO_2, broken line; HCO_3^-, full line; CO_3^{2-}, dotted line. Drawing by Mats Björk.

then it also may give the false impression that the CO_2 concentration decreases with pH (while, again, a closer look reveals that the y-axis represents relative concentrations only). In open systems where pH is changed artificially by additions of acids or bases, the dissolved CO_2 (eventually) equilibrates with the overlaying atmosphere such that it is independent of pH (i.e. if the pH is lowered, then excess CO_2 formed from HCO_3^- and CO_3^{2-} will diffuse out from the aqueous system to the atmosphere, and *vice versa* if pH is increased and the ionic carbon forms are formed from CO_2). This situation is depicted in Figure 3.5, where it can be seen that the CO_2 concentration is independent of pH, while HCO_3^- and CO_3^{2-} concentrations increase exponentially with pH. Also here, however, the Ci concentrations as a function of pH are not in agreement with those of natural systems where pH changes are based on photosynthetic and/or respiratory activities: In nature, total Ci changes as it is used up or generated when organisms photosynthesise or respire, respectively, but the total alkalinity

remains unchanged as OH^- or H^+ compensate for changes in the ionic Ci forms; this is not the case in Figure 3.5 where Ci increases at high pH (see also Box 3.4). In natural seawater, which has a pH of 8.1 in equilibrium with today's atmospheric CO_2 concentration, changes in pH usually occur as a result of plant photosynthesis and plant as well as animal respiration (Box 3.4). If so, while the ionic Ci forms (which cause the so-called carbon or carbonate alkalinity) change (concentrations decrease if photosynthesis exceeds respiration and *vice versa*), the total alkalinity remains unchanged (again since use or release of CO_2 and ionic Ci are compensated for by the formation or withdrawal of OH^- or H^+, see more about this in Chapter 7). Such natural aquatic systems in which pH varies, e.g. dielly are thus not in equilibrium with the atmosphere; if they were, then the CO_2 used or generated would equilibrate with the atmosphere such that no changes in CO_2 or pH would be noticeable. A plausible natural system in which pH changes due to photosynthesis and respiration is depicted in

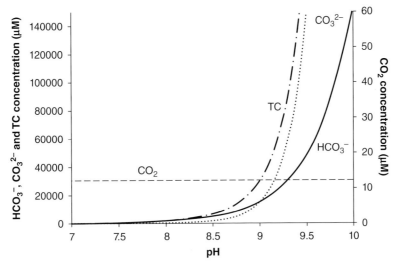

Figure 3.5   Concentrations of inorganic carbon (Ci) forms as a function of pH in an open seawater system (at 20 °C and a salinity of 35) in which pH is changed artificially by adding acids or bases, thus changing the total alkalinity. Observe that the CO_2 concentration remains the same at each pH value and is only a function of the CO_2 concentration of the atmosphere above (containing, in this case, 394 ppm or ~18 µM, representing ~13 µM dissolved CO_2). Total Ci increases with pH as both HCO_3^- and CO_3^{2-} increase. The graph does not take into consideration that high CO_3^{2-} concentrations will cause the precipitation of $CaCO_3$, which in small volumes would fill up the system. CO_2, broken line; HCO_3^-, full line; CO_3^{2-}, dotted line; Total Ci (TC), broken/dotted line. Drawing by Mats Björk.

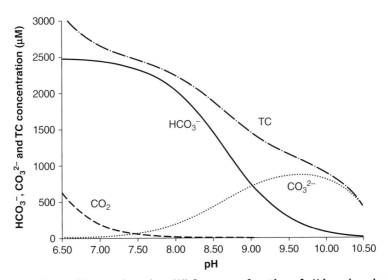

Figure 3.6   Concentrations of inorganic carbon (Ci) forms as a function of pH in a closed natural seawater system (at 20 °C and a salinity of 35) in which pH is changed by photosynthesis (towards the right) and respiration (towards the left), thus changing the carbonate alkalinity but keeping the total alkalinity constant. While HCO_3^- concentrations increase with pH up to 8, this is not apparent in the graph as the total Ci decreases towards the right due to photosynthetic activity. CO_2, broken line; HCO_3^-, full line; CO_3^{2-}, dotted line; Total Ci (TC), broken/dotted line. Drawing by Mats Björk.

Figure 3.6, where it can be seen that unlike in the two previous figures, the total Ci concentration decreases, which leads to an increase in pH (as CO_2 and, thus, H_2CO_3 is reduced). Usually, natural systems containing biota (plants and animals) exist somewhere between open (fully air-equilibrated) and closed (e.g. in calm, highly photosynthetic shallow bays, or rockpools, see Sections 11.3 and 11.2, respectively). It should be noted that all Ci forms in Figures 3.4–3.6 are dissolved. However, since calcium (Ca) is also present in aquatic systems, and especially so in seawater (at a ~10 mM concentration), the ionic Ci-form CO_3^{2-} increases its tendency to form $CaCO_3$ as pH rises, and this may precipitate out of the water as a solid. When this occurs, such as in calcification (see Box 2.3 and Section 7.5), then the total dissolved Ci will decrease, causing a lowering of pH, e.g. quite the opposite from withdrawing CO_2 by photosynthesis (see Box 2.4). The formation of $CaCO_3$ at high pH values within or outside of cell walls is the basis for calcification in, e.g. calcareous algae and corals (see Sections 7.5 and 9.3, respectively).

The reader may ask why we present three graphs of changes in the different Ci forms as a function of pH? While it is true that only Figure 3.6 represents the Ci concentrations under natural conditions when the seawater pH deviates from 8.1, Figure 3.4 is much more commonly used in textbooks. The latter is a good general representation of Ci concentrations in a relative manner, but is hard to implement for natural systems since the total Ci in the latter change and would have to be measured and then taken into consideration when calculating actual concentrations of the different Ci forms (again, since Figure 3.4 only shows relative concentrations). The data in this figure (Figure 3.4) and Figure 3.5 are, however, very useful when generating artificial systems in which short-term marine plant photosynthesis can be measured as a function of

either changing CO_2 and HCO_3^- concentrations (Figure 3.4) or changing HCO_3^- concentrations with maintained, air-equilibrated, CO_2 concentrations (Figure 3.5); examples of such applications will be given in Section 7.3. If the relative levels of the different Ci forms are sought in natural marine systems that are sufficiently closed such that pH changes can be recorded, then Figure 3.6 is the relevant one. Such situations occur, e.g. in highly productive bays where macrophytes prevail (such as will be described in Sections 11.2 and 11.3), while in open waters containing only low concentrations of phytoplankton the dissolved and atmospheric CO_2 will be in equilibrium and pH will remain constant.

For the interested reader and/or would-be experimenter, there are programs that calculate Ci concentrations at given pHs, atmospheric CO_2 concentrations, total-Ci concentrations, total alkalinities, temperatures and salinities, etc. These programs are available via the Internet: Commonly used software for estimating the parameters of the marine CO_2 system include 'CO2sys' (available at http://cdiac.ornl.gov/oceans/co2rprt.html) or SWCO2 (http://neon.otago.ac.nz/research/mfc/people/keith_hunter/software/swco2/). We (the authors) are also always willing to assist any person wishing to obtain additional information on this, as well as other related matters (i.e. matters concerning the contents of this entire book). Therefore, please feel free to email any of us for assistance.

3.2.1 pH

From the above, it is evident that there is a strong link between pH and Ci, be it total Ci or the distribution between its different forms. The pH of air-equilibrated oceanic (in the open ocean) seawater is today around 8.1, but can fluctuate closer to shore because of influences

> ## Box 3.4   Inorganic carbon equilibria, the total alkalinity and pH of seawater
>
> The **pH** of a water body can increase when, for example, a) photosynthesis exceeds respiration and, concomitantly, b) the changes in inorganic carbon (Ci) are faster than the exchange of CO_2 between the liquid and the gaseous phase overlaying it (also called the air–water inter-phase). This can easily be observed in, e.g. a well-lit aquarium where, if the plants are real and not made of plastic, pH increases as the day progresses. The increase in pH is principally due to photosynthesis using up Ci faster than both plants and fish respire it and faster than it can be replenished by the dissolution of atmospheric CO_2 (see, e.g. in Reaction 3.3). Under such conditions, Ci, including CO_2 and H_2CO_3, decrease during the day such that the acidic component decreases and, accordingly, pH increases. Conversely, during the night-time respiration causes a surplus of dissolved CO_2 that, again, results in a decrease in pH (according to Reaction 3.3). (Incidentally, pH will decrease in a plastic-plant aquarium also during daytime if it contains fish that breathe out more CO_2 than can be equilibrated with the atmosphere through the water surface; plastic plants in aquaria are, though, an insult to biologists!)
>
> The total **alkalinity** of water is determined by the sum of the bases dissolved therein. These include in seawater mainly HCO_3^-, CO_3^{2-} (forming what is termed carbonate alkalinity) and OH^- (and some borates and phosphates and other alkaline ions). If pH is increased by adding a base in a closed system, the alkalinity increases (OH^- is added) while the Ci remains constant (as long as $CaCO_3$ is not precipitated). However, if pH increases by photosynthesis such as depicted in Figure 3.6, then Ci is lowered by its photosynthetic utilisation, and since the pH increases, the total alkalinity remains constant. These are the basic differences when calculating the different Ci forms in experimental *vs.* natural systems as a function of pH. Contrary to the photosynthetic removal of CO_2, during calcification (i.e. formation of $CaCO_3$), the total Ci will decrease while pH also decreases and, thus, the total alkalinity will decrease too. As a caution, alkalinity is not the same as basicity (the opposite of acidity): The pH of water can be lowered by adding CO_2; thus its basicity will be lowered but the total alkalinity will remain the same (since the addition will increase to the carbonate alkalinity).

from land runoff and river inflows. Furthermore, metabolic activities that release or use CO_2 may influence the seawater pH in areas with high concentrations of organisms. The pH of seawater is partly due to the alkaline ions therein, and this determines the distribution of the different Ci forms that, accordingly, buffer the seawater pH to 8.1. As explained above, additions or removals of CO_2 will lower or raise the seawater pH, respectively. Thus, in equilibrium with air, increasing atmospheric concentrations of CO_2 will result in increasing concentrations of H_2CO_3 and, consequently, a lowering of the seawater pH, a phenomenon known as ocean acidification (see further in Section 12.2).

Except for changing the concentrations of the ionic Ci forms relative to CO_2, pH also affects the forms of various nutrients other than Ci. For example, high pH values promote the formation of ammonia (NH_3) from ammonium (NH_4^+), the former N-form of which can become toxic to marine plants. pH can also influence the availability of other elements, such as iron and other trace elements, to marine plants.

Table 3.1 A simple composition of synthetic
seawater containing approximately the same ratio of
elements as natural seawater at a salinity of 35. In
order to adjust the above recipe for different
salinities, the amounts of salts can be altered
accordingly while keeping their ratio constant (or by
diluting the solution with distilled water).

Salts	mM	g L^{-1}
NaCl	500	29.2
MgSO$_4$	30	3.6
CaCl$_2$	10	1.1
KCl	10	0.75
NaHCO$_3$	2	0.17

3.3 Other abiotic factors

3.3.1 Salinity

The most striking difference between fresh
waters and seawater is that the latter contains high amounts of salts. Mid-oceanic waters
contain 3.1–3.8% salts, mostly composed of
sodium (Na$^+$) and chloride (Cl$^-$) ions (see
Table 3.1 for a simplified composition of seawater). More commonly, such salinities are
expressed on a per mill (‰) basis such that the
above concentrations equal 31–38‰ (or g kg^{-1}
of seawater), while coastal waters have a much
lower salinity (depending on the contribution of freshwater from rivers and rains). The
salt concentrations of seawaters have classically
been derived by evaporating the water from 1
kg of seawater, and then weighing the remaining salts. Nowadays, however, salinity is usually measured by a (appropriately calibrated)
conductivity meter, using the practical salinity
scale (PSS), where the salt content is derived
from the ratio between the conductivity of a
water sample and a standard KCl solution and,
therefore, has no units to it. In practice, however, a salinity of say 35 corresponds to a salt
content of approximately 35 g kg^{-1} (i.e. 35‰).

Seawater is composed of almost all elements present in the Periodic Table. Major ions
other than Na$^+$ and Cl$^-$ are those of magnesium (Mg^{2+}), sulphate (SO$_4^{2-}$), calcium (Ca^{2+})
and potassium (K$^+$). In addition to the ele-
ment C (in the appearance of various inorganic carbon, or Ci, forms, see above), N, P
and Fe (see below) are especially important for
the metabolism of plants in general (including the marine ones) and silica (Si) is essential for the diatoms, and may be limiting to
them in seawater, while other micronutrients
such as cobalt (Co), boron (B) and copper (Cu)
usually are present at high enough concentrations to sustain healthy plant growth. Marine
plants often store nutrients such as N and P,
and it is thus possible to keep plants active in
synthetic seawater solutions for periods of days
without adding those nutrients. We have successfully used the simple composition of synthetic seawater as given in Table 3.1 for photosynthetic experiments on macroalgae and seagrasses. It is interesting to note, e.g. in that
recipe, that Na and Cl, which are present in seawater at much higher concentrations than any
other elements or ions, are not used much for
metabolic processes of plant cells such as photosynthesis and respiration, while, e.g. HCO$_3^-$
and ions of lower concentrations are of great
importance.

For photosynthesis, there are two major
effects of salts: First, salty waters usually contain less dissolved CO$_2$ than freshwater at a
given temperature (even though the total concentration of Ci is usually higher in saltwater
because of the higher concentration of ionic
carbon forms therein, which in turn is due
to its, usually, higher pH), and this may have
effects on the acquisition of inorganic carbon (Ci). Secondly, salts negatively affect the
activity of enzymes of, e.g. the Calvin cycle
of CO$_2$-fixation and -reduction (see Chapter 6), and must therefore be lower within the
cells and their chloroplasts than in highly (or

even normally) saline seawaters. This is usually accomplished by restricting the inflow of ions at the plasma membrane level, but can also be due to the efficient active removal of specific salts that penetrate into the cells, resulting in lower intracellular concentrations than in the surrounding seawater. Using this latter means of salt restriction it was for example estimated that the Na and Cl concentrations in the chloroplast-containing cells of some seagrasses were 120 and 80 mM, respectively, as compared with about 500 mM of both ions in the surrounding seawater; those intracellular concentrations affected the activity of the key photosynthetic enzyme ribulose-1,5-bisphosphate carboxylase/oxygenase (Rubisco) much less than did the salinity of seawater (see the ancient paper by Beer, Eshel and Waisel, 1980). A third way to control the intracellular salt concentration is to form compatible organic solutes such as proline and/or sugars that do not affect biochemical processes but keep up the internal osmotic pressure such that salt ions do not enter the cells easily. In many marine algae, the mechanisms of lowering the concentration of salts within the cells are so effective that the internal salt concentrations are kept at very low levels and, so, do not affect metabolism.

3.3.2 Nutrients

Except for Ci (CO_2 and/or HCO_3^-, see Chapter 7), marine plants also need other nutrients in order to grow. Classically notable among those nutrients that may **limit photosynthesis and growth** for marine and freshwater plants are **nitrogen** (N) and **phosphorus** (P), respectively, but it is increasingly recognised that P may be limiting in marine environments too. Levels of those nutrients vary greatly between different marine habitats: Total concentrations of ionic N, i.e. nitrate (NO_3^-), nitrite (NO_2^-) and ammonium (NH_4^+), can be found at around 50–100 μM but can be much lower in nutrient-poor mid-oceanic, and much higher in nutrient-rich, near-shore, waters. The concentration of dissolved nitrogen gas, N_2, is higher but only few organisms (i.e. some cyanobacteria) can use this N-form for their nutritional (or metabolic) needs (see Box 2.1). Phosphorus (in the form of PO_4^{3-}) concentrations are typically lower, i.e. some 5 μM, but, like N, fluctuate in different water types. A rule seems to be that the molar N:P ratio is rather stable at 15 irrespective of water type.

Oceanic surface waters in particular may contain low concentrations of N and P since these nutrients are constantly being used up by photosynthesising organisms within the photic zone. Such biological activities by marine plants can cause the depletion of other crucial elements as well, though coastal waters are less frequently nutrient limited than the open oceans. Intense growth of diatoms may deplete seawater of silica (Si), causing severe limitations on further diatom growth and resulting in these species being succeeded temporally by species that do not require Si. More importantly, the primary productivity of phytoplankton in large areas of the Southern Ocean and the sub-equatorial Pacific have been shown to be limited by iron (Fe), and even zinc (Zn) has been implied as a limiting nutrient in some situations. Iron limitation has been demonstrated in large-scale ocean enrichment experiments in which added dissolved Fe was shown to stimulate phytoplankton growth and photosynthesis. At one stage these Fe-enrichment experiments were touted as a potential remedy for ameliorating global CO_2 increases, but it is now recognised that this approach is untenable since Fe additions are very short-lived and can potentially alter ecological processes in the water column in an unpredictable way. (See more about the potential use of Fe (and other nutrient) enrichments to increase phytoplankton productivity in Section 12.1.)

3.3.3 Temperature

Temperature affects the solubility of gases such as CO_2 in seawater equilibrated with air (see Section 2.2) and, to a lesser extent, the equilibrium distribution between CO_2 and the ionic forms HCO_3^- and CO_3^{2-}. Thus, temperature may (potentially) have an indirect effect on marine plants by changing the concentrations of Ci-forms used in photosynthesis. Temperature also affects the diffusion coefficient of gases, values for CO_2 being 1.92×10^{-9} m^2 s^{-1} at 25 °C and 1.30×10^{-9} m^2 s^{-1} at 5 °C. Furthermore, there are more direct impacts of temperature on biochemistry (including the kinetic characteristics of the central CO_2-fixing enzyme in photosynthesis, ribulose-1,5-bisphosphate carboxylase/oxygenase, Rubisco; see more about that important enzyme in Chapter 6) and metabolic rates, and hence on growth rates, of marine plants.

Although marine plants can be found in a wide range of temperature regimes, from the tropics to polar regions, the large bodies of water that are the environment for most marine plants have relatively constant temperatures, at least on a day-to-day basis. This is because the specific heat of water is some four times that of air on a mass basis (J kg^{-1} K^{-1}; K = Kelvin) or 4000 times that of air on a volume basis (J L^{-1} K^{-1}). For these plants, therefore, temperature differences in the marine environment are mainly associated with seasonal and latitudinal variations. For marine plants that are found in intertidal regions, however, temperature variation during a single day can be very high as the plants find themselves alternately exposed to air (which can be very much hotter or colder than the seawater, depending on location and season).

Marine plants from tropical and temperate regions tend to have distinct temperature ranges for growth (which are, however, frequently overlapping) and growth optima. For example, among most temperate species of microalgae, temperature optima for growth are in the range 18–25 °C, while some Antarctic diatoms show optima at 4–6 °C with no growth above a critical temperature of 7–12 °C. By contrast, some tropical diatoms will not grow below 15–17 °C. Similar responses are found in macroalgae and seagrasses. However, although some marine plants have a restricted temperature range for growth (so-called stenothermal species; steno = narrow and thermal relates to temperature), most show some growth over a broad range of temperatures and can be considered eurythermal (eury = wide). For microalgal growth in the range of maximum sensitivity to temperature, growth rate (μ) is related to temperature (T) by the equation $\mu = \alpha(T - c)^\beta$ where α and β are constants and c is the minimum temperature limit of growth (see Ahlgren, 1987).

A useful indicator of the sensitivity of metabolic processes and growth to temperature is the parameter Q_{10}, which is a measure of the proportional rise in the rate process with a 10 °C rise in temperature (Q_{10} = rate $_{(t+10\,°C)}$/ rate$_{(t\,°C)}$). In photosynthesis, purely photochemical processes such as light absorption and transfer of excitation energy (see Section 5.1) are essentially temperature independent ($Q_{10} = 1$). However, CO_2 fixation (Section 6.1) and, to a lesser extent, thylakoid reactions involving electron transport have Q_{10} values more typical of biochemical reactions, which vary from ~ 2.0 to 2.6 (see the review by J.A. Raven and R.J. Geider, 1988).

3.3.4 Water velocities

Water velocities are central to marine photosynthetic organisms because they affect the transport of nutrients such as Ci towards the photosynthesising cells, as well as the removal of by-products such as excess O_2 during the day. Such bulk transport is especially important in aquatic media since diffusion rates there are typically some 10 000 times lower than in air (see discussion on diffusion boundary layers in Chapter 7). It has been established that increasing current velocities will increase photosynthetic rates and, thus, productivity of macrophytes as long as they do not disrupt the thalli of macroalgae or the leaves of seagrasses. Typical current velocities that cause enough transport of Ci and nutrients such as not to limit photosynthesis and growth severely in both macroalgae and seagrasses are 1–20 cm s^{-1}. It is harder to determine current requirements for planktonic cyanobacteria and microalgae since they are carried with the currents and, thus, experience a lower relative flow rate surrounding them than the rate of the current. Furthermore, given the size of such organisms, boundary layer thicknesses are considerably smaller and have less impact on diffusion resistance.

Summary notes of Part I

- **Life on Earth** may have started in ponds near volcanoes, or in water droplets, by combining non-oxidising gases as energised by, e.g. lightning, some 3.5 billion years ago. Life may, however, also have started elsewhere in the universe, and brought to Earth by, e.g. meteorites from other planets in our solar system.
- **Life evolved in the oceans** for over 3 billion years, where eukaryotic cells evolved from prokaryotic ones some 1.5 billion years ago through the process of endosymbiosis. Plants then invaded land some 400 million years ago.
- Marine 'plants' (= all photoautotrophic organisms of the seas) can be divided into **phytoplankton** ('drifters', mostly unicellular) and **phytobenthos** (connected to the bottom, mostly multicellular/macroscopic).

The phytoplankton can be divided into **cyanobacteria** (prokaryotic) and **microalgae** (eukaryotic); some of these microscopic forms live as **photosymbionts** in many marine invertebrates. The phytobenthos can be divided into **macroalgae** and **seagrasses** (marine angiosperms, which invaded the shallow seas some 90 million years ago). The micro- and macro-algae are divided into larger groups as based largely on their pigment composition.

- There are some 150 currently recognised species of marine cyanobacteria, ~20 000 species of eukaryotic microalgae, several thousand species of macroalgae and 50(!) species of seagrasses. Altogether these **marine plants are accountable for approximately half of Earth's photosynthetic (or primary) production**.
- The **abiotic factors** that are conducive to photosynthesis and plant growth in the marine environment differ from those of terrestrial environments mainly with regard to light and inorganic carbon (Ci) sources. **Light** is strongly attenuated in the marine environment by absorption and scatter, and may limit photosynthesis and growth at irradiances that are 0.1–10% (with 1% as a possible average) of the surface irradiance. Light for photosynthesis is measured as a flux of photons within the range of 400 nm (blue) to 700 nm (red), which is thus called the **photosynthetically active radiation** (**PAR**); the irradiance (or 'light intensity') on a **sunny day is ~ 2000 µmol photons m^{-2} s^{-1}**. While terrestrial plants rely of atmospheric CO_2 for their photosynthesis, marine plants utilise largely the >100 times higher concentration of **HCO_3^- as the main Ci source for their photosynthetic needs**. Nutrients other than CO_2 that may limit plant growth in the marine environment include nitrogen (N), phosphorus (P), iron (Fe) and, for the diatoms, silica (Si).

Part II

Mechanisms of Photosynthesis, and Carbon Acquisition in Marine Plants

Introduction to Part II

Photosynthesis is the process by which the energy of light is used in order to form energy-rich organic compounds from low-energy inorganic compounds. In doing so, electrons from water (H_2O) reduce carbon dioxide (CO_2) to carbohydrates. The simplified formula for photosynthesis is:

$$CO_2 + H_2O \overset{\text{sunlight}}{\rightarrow} (CH_2O) + O_2 \quad \text{(II.1)}$$

where (CH_2O) represents the high-energy containing carbohydrate (or saccharide or, simply, sugar). If we start with 6 molecules of CO_2 and 6 of H_2O then the product reads $C_6H_{12}O_6$, which is the familiar sugar glucose. In the formula, it may at first glance appear as if O_2 (molecular oxygen) is derived from CO_2, but we shall learn in Chapter 5 that it is 'split' off from H_2O. Further, since O_2 is an end product,

it is logical that its accumulation should inhibit the entire process, which we will learn in Section 6.1 is the case for most terrestrial plants, but usually not for marine ones (see Chapter 7). Also, looking at the above formula we recognise that by reversing the horizontal arrow we arrive at the process of respiration, and in later chapters of Part III we shall also discuss the two processes, photosynthesis and respiration, with regard to one another.

The process of photosynthesis can conveniently be separated into two parts: the '**photo**' part in which light energy is converted into chemical energy bound in the molecule ATP and reducing power is formed as NADPH, and the '**synthesis**' part in which that ATP and NADPH are used in order to reduce CO_2 to sugars (see Figure II.1). The 'photo' part of photosynthesis is, for obvious reasons, also called its light reactions while the 'synthesis' part can be termed CO_2-fixation and -reduction, or the Calvin cycle after one of its discoverers; this

Photosynthesis in the Marine Environment, First Edition. Sven Beer, Mats Björk and John Beardall.
© 2014 John Wiley & Sons, Ltd. Published 2014 by John Wiley & Sons, Ltd.
Companion Website: www.wiley.com/go/beer/photosynthesis

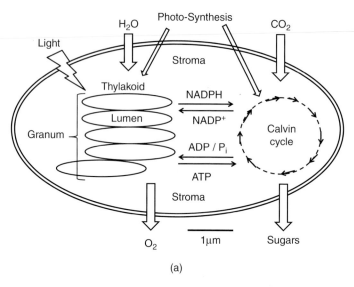

(a)

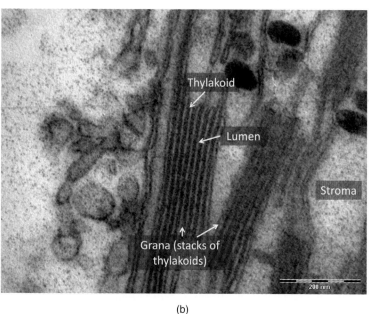

(b)

Figure II.1   (a) **A schematic drawing of a 'typical chloroplast'** and the main substrates and products of the light-reactions ('Photo'- in Photo-Synthesis, associated with the thylakoid membranes) and the CO_2-fixation and -reduction reactions, also called the Calvin cycle (-'Synthesis' in Photo-Synthesis, in the stroma). ATP and NADPH are the energy source and reducing power, respectively, formed by the light reactions, that are subsequently used in order to reduce carbon dioxide (CO_2) to sugars (synonymous with carbohydrates) in the Calvin cycle. Molecular oxygen (O_2) is formed as a by-product of photosynthesis. Note the size bar, indicating that the size of a chloroplast is typically some 5–10 μm (1 μm = 10^{-3} mm). (b) **An electron micrograph of part of a chloroplast** featuring thylakoids and other sub-organellar constituents. Note the size bar, in which each division is 0.2 μm. Drawing by Sven Beer (a), Photo with permission from, and thanks to, Bela Hausmann, University of Vienna, Austria (b).

part also used to be called the 'dark reactions' of photosynthesis because it can proceed *in vitro* (= outside the living cell, e.g. in a test-tube) in darkness provided that ATP and NADPH are added artificially. However, *in vivo* (= in the living cell) CO_2-fixation and -reduction cannot proceed because, following darkening, the cellular compartments where photosynthesis is carried out would quickly run out of ATP and NADPH, which are formed naturally only in the light.

In photosynthetic bacteria (such as the cyanobacteria), the light reactions are located at the **plasma membrane** and internal membranes derived as invaginations of the plasma membrane. Although other reactions of the CO_2-fixation and -reduction process take place in the soluble phase of the cytoplasm, most of the CO_2-fixing enzyme ribulose-bisphosphate carboxylase/oxygenase (Rubisco, see Chapter 6) is here located in structures termed **carboxysomes**. Carboxysomes are polyhedral bodies with a semi-permeable protein coat surrounding Rubisco and another enzyme important in the photosynthetic apparatus of marine plants, carbonic anhydrase (CA, see Chapter 7), and are found in many chemoautotrophs[1] as well as in **photoautotrophs**[1] such as the cyanobacteria (see below). In all other plants (including algae), however, the entire process of photosynthesis takes place within intracellular compartments called **chloroplasts** which, as the name suggests, are chlorophyll-containing plastids (plastids are those compartments in cells that are associated with photosynthesis). A 'typical' chloroplast is illustrated in Figure II.1a. In its basic form it is typically surrounded

by a double membrane (each one of which, like other cellular membranes, is comprised of a lipid bi-layer; remember also from Section 2.2 that the chloroplasts of some groups such as diatoms and dinoflagellates have different numbers of bounding membranes reflecting their different endosymbiotic origins) and contains regions of intra-chloroplastic membranes forming flattened sack-like vesicles termed **thylakoids**, which are surrounded by a membrane-free fluid termed the **stroma**. (Also the internal photosynthetic membranes of the cyanobacteria are called thylakoids.) Further, the thylakoids can be found stacked on top of one another to form structures called **grana** (granum in singular, = grain). The thylakoids have an inner fluid part called the **lumen**, while their outside either borders another thylakoid or is exposed to the stroma. While the light reactions occur within, or in close proximity to, the stromal membranes (see Chapter 5), the CO_2-fixation and -reduction reactions are located in the stroma (see Chapter 6). A 'real' picture of part of a chloroplast can be seen in Figure II.1b. In some species the majority of the Rubisco is localised to a structure within the chloroplast termed the **pyrenoid**, which is analogous to the carboxysome in autotrophic bacteria (although it lacks a protein coat or bounding membrane); in the 'typical' case, though, Rubisco, as well as the other parts of CO_2-fixation and -reduction, is located in the stroma.

The chloroplast-containing cells of the green parts (mainly leaves) of higher plants typically contain tens to hundreds of chloroplasts. Some green algae contain only one (albeit huge) chloroplast per cell (e.g. *Ulva* spp., see Figure 2.4b) while the structure and colour of chloroplasts of other algae have in the past rendered them names such as rhodoplasts (in red algae, see the next chapter). Seagrasses, which are descendants of terrestrial higher plants, naturally contain chloroplasts typical of the latter.

---

[1] Remember, chemoautotrophs convert inorganic carbon (Ci) to organic material using energy from the oxidation of reduced forms of elements, e.g. NH_4^+ (ammonium) and H_2S (hydrogen sulphide). Photoautotrophs, on the other hand, use sunlight as the energy source for this process. (This book is about photoautotrophs!)

The process of photosynthesis has evolved in marine plants over 3.4 billion years while terrestrial, 'higher', plants became established only during the past *ca.* 350 million years (see Chapter 1). Therefore, and not surprisingly, both groups of plants feature similar basic mechanisms of photosynthesis on the cellular level. Today (in evolutionary terms, i.e. during the past few 100 millions of years), however, CO_2 concentrations in both the atmosphere and seawater have dropped to values where this substrate would potentially restrict photosynthesis and growth in both the terrestrial and, at least well-lit, marine environments: in marine systems the low CO_2 concentration is exacerbated by diffusivity constraints and in terrestrial environments by the potential lack of water. In marine systems, however, CO_2 equilibrates with other inorganic carbon (Ci) forms such that at today's seawater pH (8.1) the concentration of bicarbonate (HCO_3^-) is >100 times that of CO_2. These latter differences between the marine and terrestrial environments have, in our view, led to the differences in photosynthetic 'strategies' between the two plant groups, as stressed in this book: Most marine plants possess the capability of utilising HCO_3^- as a way to concentrate CO_2 for their intracellular photosynthetic needs (see Chapter 7), while some terrestrial plants have, especially in arid environments, found other ways to concentrate CO_2 so as to optimise photosynthesis in a low-CO_2 atmosphere (see Section 6.2). These differences in Ci acquisition will hopefully (read: certainly) become clearer later in this part of the book.

Chapter 4

Harvesting of light in marine plants: The photosynthetic pigments

Photosynthesis can be seen as a process in which part of the radiant energy from sunlight is 'harvested' by plants in order to supply chemical energy for growth. The first step in such light harvesting is the absorption of photons by photosynthetic **pigments**[1]. The **photosynthetic pigments** are special in that they not only convert the energy of absorbed photons to heat (as do most other pigments), but largely convert photon energy into a flow of electrons; the latter is ultimately used to provide chemical energy to reduce CO_2 to carbo-

hydrates. This chapter will introduce the reader to the various photosynthetic pigments residing in all plants (e.g. chlorophyll a and certain carotenoids), and pigments that are special for specific marine plants (e.g. phycobilins in cyanobacteria and red algae). The way that plants utilise the light in photosynthesis will then be the subject of Chapters 5–7.

4.1 Chlorophylls

As stated above, the photosynthetic membranes contained within cyanobacterial cells and the thylakoid membranes of algal and seagrass chloroplasts are the sites of the light-reactions in photosynthesising marine plants. Since absorption of photons by the photosynthetic pigments constitutes the initial step in photosynthesis, it follows that those pigments are located to those membranes. Chlorophyll is a major photosynthetic pigment, and

---

[1]Pigments are substances that can absorb different wavelengths selectively and so appear as the colour of those photons that are less well absorbed (and, therefore, are reflected, or transmitted, back to our eyes). (An object is black if all photons are absorbed, and white if none are absorbed.) In plants and animals, the pigment molecules within the cells and their organelles thus give them certain colours. The green colour of many plant parts is due to the selective absorption of chlorophylls (see below), while other substances give colour to, e.g. flowers or fruits.

Photosynthesis in the Marine Environment, First Edition. Sven Beer, Mats Björk and John Beardall.
© 2014 John Wiley & Sons, Ltd. Published 2014 by John Wiley & Sons, Ltd.
Companion Website: www.wiley.com/go/beer/photosynthesis

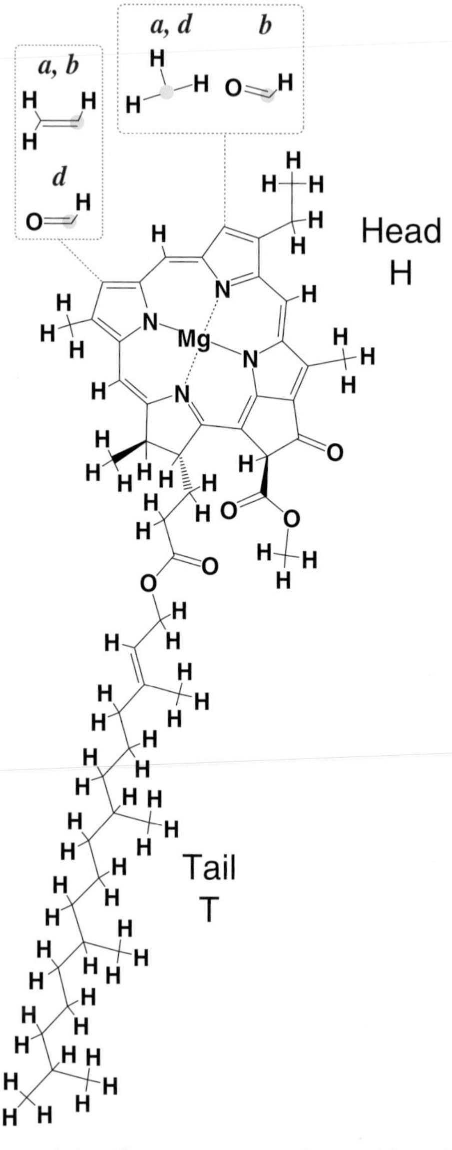

Figure 4.1 The structure of a chlorophyll molecule depicting its hydrophilic 'head' (H) and hydrophobic 'tail' (T), as well as the differences in chlorophyll *a*, *b* and *d* (within dotted boxes). From Wikipedia Commons (free to be used by any one), http://commons.wikimedia.org/wiki/File:Chlorophyll _structure.png.

repelling) molecule[2] with a molecular weight of about 900. One part of the molecule (the 'head') is comprised of four pyrrole rings held together by a magnesium (Mg) atom to form a larger porphyrin-like ring structure, while a hydrocarbon chain forms the 'tail'; it is the latter that makes the molecule hydrophobic. On the other hand, the tail makes the molecule lipophilic (lipid attracting), and this feature anchors the chlorophyll in the (lipid-based) thylakoid membranes. The molecular sub-structure of the chlorophyll's 'head' makes it absorb mainly blue and red light (see Figure 4.2), while green photons are hardly absorbed but, rather, reflected back to our eyes (and, in some thin leaves or algal thalli, also penetrated through the plant bodies and seen by our eyes on their other side) so that chlorophyll-containing plant parts look green. More importantly, as for all photosynthetic pigments, the molecular structure of chlorophyll makes it amenable to become excited (= transfer electrons to higher energy levels, read: orbitals further away from the nucleus) by the photons that it absorbs, and the subsequent de-excitation of chlorophyll (as well as other photosynthetic pigments) as the electrons subsequently return to their original level ultimately gives rise to electron flow in photosynthesis (see Chapter 5).

As can be seen in Figure 4.1, chlorophylls *a* (present in all plants), *b* (present in higher

chlorophyll *a* is present in all plants, including all algae and the cyanobacteria. The molecular structure of chlorophyll is depicted in Figure 4.1. It is a largely hydrophobic (water-

[2]The fact that chlorophyll is hydrophobic can easily be seen if trying to extract it in water; even when boiling, e.g. a green leaf such as spinach, the leaf remains green while the water remains virtually colourless. Chlorophyll can, on the other hand, easily be extracted into organic solvents such as acetone, methanol or dimethyl formamide (DMF) for experimental uses, or various oils (e.g. in cooking if a green colour of the non-leafy ingredients is desired; it has been said that in past times meat-containing green vegetable casseroles, when cooked in plenty of oil, caused not only the oil but also the meat to become green, but this needs yet to be confirmed, experimentally, in the kitchen!).

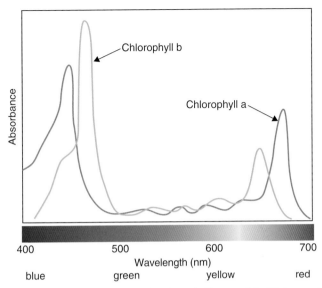

Figure 4.2   Absorption (Absorbance) **spectrum of chlorophylls *a* and *b*.** With permission from, and thanks to, John Tiftickjian, Delta State University, Cleveland, USA.

plants and green algae), *c* (present in the diatoms, dinoflagellates and brown macroalgae) and *d* (present in the red algae) have very similar structures. Yet, the slight differences in the head parts of the molecules confer on the different chlorophylls somewhat different absorption spectra in the blue- and red-absorbing range (see, e.g. the difference in absorption between chlorophylls *a* and *b* in Figure 4.2). Thus, e.g. chlorophylls *b*, *c* and *d* have absorption maxima at slightly higher wavelengths in the blue part of the spectrum and at lower ones in their red absorption region. Chlorophyll *d* has an additional peak, at least when extracted into an organic solvent, in the far-red wavelength end of the spectrum (see Table 4.1).

4.2   Carotenoids

In addition to chlorophyll *a*, all plants contain carotenoids. Some, like β-carotene, are ubiquitous and are active in light harvesting in pho-

tosynthesis in seagrasses, algae and cyanobacteria. Other carotenoids such as fucoxanthin, peridinin and 19′-butanoyloxyfucoxanthin are found only in particular groups of algae (see Table 4.2). Fucoxanthin is the major carotenoid in the Chromophyte algae (comprising the brown algae, diatoms, haptophytes and cryptomonads). It is also found in some dinoflagellates, though predominantly these use peridinin. 19′-butanoyloxyfucoxanthin is a characteristic marker for the Pelagophyceae and 19′-hexanoyloxyfucoxanthin is used in

Table 4.1   Absorption peaks of chlorophylls in 90% acetone: the middle column shows absorption in the blue band and the right column absorption in the red band.

Pigment	Absorption peaks	
chlorophyll *a*	430 nm	664 nm
chlorophyll *b*	460 nm	647 nm
chlorophyll *c1*	442 nm	630 nm
chlorophyll *c2*	444 nm	630 nm
chlorophyll *d*	401 nm and 455 nm	696 nm

Table 4.2   Major pigments involved in photosynthetic light harvesting in algae (after Gantt and Cunningham, 2001) and seagrasses.

Pigment/Group	Cyanobacteria	Prochlorophytes	Green algae	Red algae	Chromophytes	Dinoflagellates	Seagrasses
Chl a	✓	✓[1]	✓	✓	✓	✓	✓
Chl b		✓[1]	✓				✓
Chl c_1, c_2[5]					✓	✓	
Chl d		✓[4]		✓			
Phycobilins[6]	apc, pc, pe				pc, pe[2]		
Main carotenoids	β-carotene[3]	β-carotene	β-carotene	β-carotene	β-carotene fucoxathin	β-carotene peridinin[8]	β-carotene
Other important xanthophylls	myxoxanthin		lutein	lutein	alloxanthin[7]		lutein

[1]Prochlorophytes (whereof *Prochlorococcus* is a member) contain divinyl chlorophylls a & b instead of, or in addition to (depending on species), chlorophylls a and b

[2]Found in Cryptophyceae

[3]β–carotene is found in the reaction centres of most cyanobacteria and algae (as well as higher plants)

[4]To date found only in *Acaryochloris*

[5]Most Chromophytes have c_1 and c_2 but the Cryptophyceae have only chl c_2, as do the red algae

[6]apc = allophycocyanin; pc = phycocyanin; pe = phycoerythrin

[7]In the cryptomonads

[8]Some dinoflagellates have fucoxanthin rather than peridinin as their major carotenoid

Chromophytes = brown algae + diatoms

Dinophytes = dinoflagellates

coccolithophorids. All these accessory pigments act to fill in the 'green window' generated by the chlorophylls' non-absorbance in that band (see Figure 4.2) and, thus, broaden the spectrum of light that can be utilized by these algae for photosynthesis beyond that absorbed by chlorophyll.

4.3   Phycobilins

In red algae and cyanobacteria, the main pigments of the 'photosynthetic units' (see Section 5.1) are the phycobiliproteins (phycoerythrin, phycocyanin and allophycocyanin) rather than chlorophyll, and these proteins are borne externally (as extrinsic complexes) to the thylakoid membrane as structures termed phycobilisomes, protruding into the stroma in red algae or into the cytosol in cyanobacteria, rather than being embedded within the membrane (termed intrinsic complexes). The organisation of these systems is illustrated in Figure 4.3. In cryptomonads, the phycobiliproteins are not organised into extrinsic phycobilisomes, but instead are found as rod-like structures within the thylakoid lumen. Phycobilins are the primary light-harvesting pigments in the algae in which they are present, with light energy being eventually channelled to chlorophyll anchored in the thylakoid membranes as, in the case depicted in Figure 4.3, part of PSII.

It was long thought that the red colour of red algae conferred on them an advantage when growing in deep coastal waters where, because of absorption by the water molecules and scatter by particles, green light prevails (see Section 3.1). From this thought developed the theory of '**complementary chromatic adaptation**': Since green is a complementary colour of red, it was thought that the red pigment could use this green light and, thus, algae containing it could penetrate deeper than algae of different colours. Thus, the spectral composition of the submarine light field was seen as

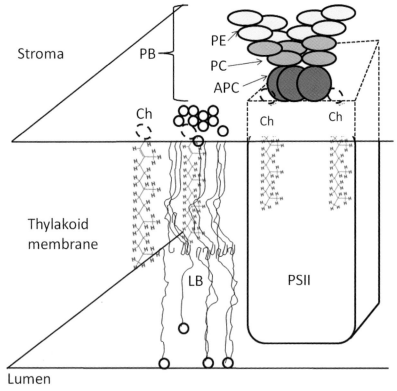

Figure 4.3   Organisation of a phycobilisome (PB) 'on top' of photosystem II (PSII) embedded into the lipid bi-layer (LB, the lipid molecules of which are indicated by small circles with tails) of a thylakoid membrane. The phycobilisome comprises molecules of phycoerythrin (PE), phycocyanin (PC) and allo-phycocyanin (APC); shown are also some chlorophyll molecules (Ch) anchored in the thylakoid membrane. Drawing by Sven Beer.

a mechanism structuring macroalgal communities. However, later observations could not confirm that red algae necessarily grew deeper than other algae. Rather, in most cases it was observed that some thick-thallied green algae, e.g. of the genera *Codium* or *Udotea*, grew well in deep waters. Elegant experiments by Matt Dring and Joseph Ramus in the early 1980s showed that adaptation to light intensity (or irradiance) rather than spectral quality is what governs the success of red algae at great depth. Thus, the theory of complementary chromatic adaptation was finally put to rest in the 1980s, e.g. in a nice article entitled "New Light on Seaweeds" by Mary Beth Saffo.

The photosynthetic pigments are essential in transferring light energy towards the reduction of CO_2 to carbohydrates, as will now be described in the next two chapters. This is, however, by no means an extensive treatise on the mechanisms of photosynthesis, and only the basic principles will be given here as a background to the subsequent detailed treatment of photosynthesis in the marine environment (in Chapter 7 and onwards). Interested readers are referred to more general texts like Plant Physiology by Lincoln Taiz and Eduardo Zeiger or the more specific book on photosynthesis entitled Molecular Mechanisms of Photosynthesis by Robert Blankenship.

The brown alga *Chorda* and filamentous algae at the Swedish west coast. Photo by Katrin Österlund.

Chapter 5

Light reactions

The light reactions of photosynthesis are those reactions in which the photon energy of light is converted into the chemical energy (ATP) and reducing power (NADPH) needed to reduce CO_2 to sugars in the subsequent CO_2-fixation and -reduction processes (Chapter 6). We will in this chapter divide the light reactions into: a) Light absorption by the photosynthetic pigments, their consequent excitation and de-excitation, the channelling of energy towards the reaction centres of photosystems II (PSII) and I (PSI), and the charge separation and transfer of electrons by the reactions centre chlorophylls; b) The transport of electrons from water in PSII to, eventually, ferredoxin in PSI, via the 'Z-scheme' in the thylakoid membranes, resulting in the formation of NADPH; and c) The formation of ATP through a proton motive force that 'drives' protons through an ATP synthase. Finally, in Section 5.4, we will treat other, non-'linear' and non-'synthetic', pathways of electron transport in photo 'synthesis'.

5.1 Photochemistry: excitation, de-excitation, energy transfer and primary electron transfer

Photosynthesis is principally a redox process in which carbon dioxide (CO_2) is reduced to carbohydrates (or, in a shorter word, sugars) by electrons derived from water. (See Box 5.1 regarding redox reactions.) However, since water has an energy level (or redox potential) that is much lower than that of sugar, or, more precisely, than that of the compound that finally reduces CO_2 to sugars (i.e. NADPH), it follows that energy must be expended in the process; this energy stems from the photons of **light.**

When a photon is absorbed by a photosynthetic-pigment molecule (e.g. chlorophyll, which we will use here as the principal one), one of the latter's electrons is moved to

Photosynthesis in the Marine Environment, First Edition. Sven Beer, Mats Björk and John Beardall.
© 2014 John Wiley & Sons, Ltd. Published 2014 by John Wiley & Sons, Ltd.
Companion Website: www.wiley.com/go/beer/photosynthesis

Box 5.1 Redox reactions in photosynthesis

Redox reactions are those reactions in which one compound, B, becomes reduced by receiving electrons from another compound, A, the latter then becomes oxidised by donating the electrons to B. The reduction of B can only occur if the electron-donating compound A has a higher energy level, or, in our present jargon, has a redox potential that is higher, or more negative in terms of electron volts, than that of compound B. The **redox potential**, or reduction potential, E_h, can thus be seen as a measure of the ease by which a compound can become reduced; the lower (more positive) the potential the easier it will receive electrons and become reduced, or, reciprocally, the higher (more negative) the potential, the easier a compound can give off an electron and, thus, become oxidised). Accordingly, the greater the **difference in redox potential** between compounds B and A, the greater the tendency that B will be reduced by A. In photosynthesis, the redox potential of the compound that finally reduces CO_2, i.e. NADPH, is more negative than that from which the electrons for this reduction stems, i.e. H_2O, and the entire process can therefore not occur spontaneously. Instead, light energy is used in order to boost electrons from H_2O through intermediary compounds to such high redox potentials that they can, eventually, be used for CO_2 reduction. In essence, then, the light reactions of photosynthesis describe how photon energy is used to boost electrons from H_2O to an energy level (or redox potential) high (or negative) enough to reduce CO_2 to sugars.

an orbital farther away from the nucleus of one of its atoms, whereby the molecule becomes energised (or excited). One can envision such **excitation** as the photon 'pushing', or 'knocking' (since the process is very fast), the electron 'uphill' (energetically) to an outer orbital, thus transferring its energy to the subsequently excited molecule. When in this excited state, however, the chlorophyll molecule is unstable, and the electron will very quickly (within a fraction of a second) spontaneously 'fall back' (return) to its original orbit (the ground state); when doing so, in the process called **de-excitation**, the energy released will largely be used for driving photosynthesis, while a much smaller part is diverted to generating heat and fluorescence (see below and Figure 5.1).

While high-energy photons (of short wavelengths, e.g. blue photons, see Section 3.1) knock chlorophyll-molecule electrons to a higher energetic level than, e.g. red, lower-energy, photons (see Figure 5.1), energy that causes fluorescence, energy transfer (between

pigment molecules) and primary electron transfer (or 'charge separation' at the reaction centres) is released only from the red-photon excitation level (which can also be called the low-excited state). The excess electron energy released through the de-excitation at higher than red-photon energy levels caused by the absorption of photons of different colours (blue yielding the highest electron energy level) is, on the way to the red-photon excitation level, dissipated as heat. The fact that photosynthesis occurs from one critical energy-dissipating excitation level only (i.e. again, the red-photon excitation level), is the basis for the past observations and nowadays proof that the **quantum yield** of the overall photosynthetic process is virtually the same whether it is caused by high-energy blue photons, low-energy red photons, or photons of those intermediary energies, that **are absorbed** by the photosynthetic pigments (see Box 5.2 on quantum yields).

As mentioned, it is only the energy released from the critical red-photon excitation level

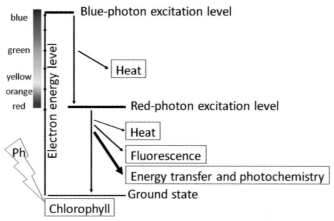

Figure 5.1   Excitation and de-excitation of chlorophyll. As photons (Ph) are absorbed by chlorophyll molecules, the latter are said to become excited as electrons are knocked towards higher energy levels (upwards arrows). The level of excitation depends on the photon energy such that blue (high-energy) photons cause the highest electron energy ("blue electron excitation"), red (low-energy) photons the lowest ("red electron excitation") and photons of other wavelengths having intermediary energy levels. Nanoseconds (ns, $= 10^{-9}$ s) after excitation, the electrons spontaneously (through various intermediate energy levels depending on the previous level of excitation) return to their normal orbits (ground state) as the molecules undergo de-excitation (downward arrows). The energy released during de-excitation towards the red-photon excitation level generates heat (with the higher energy level the electrons had reached, the more heat being released). Transfer from the red-photon excitation level towards the ground state also results in a small portion of the released energy generating heat but, unlike transfer from higher electron energy levels, some of the energy is converted to fluorescence, while most of the energy causes energy transfer (through resonance) to neighbouring pigment molecules and, eventually, primary electron transfer (also called photochemistry, at the reaction centres, see text) as the initial steps in photosynthesis. Drawing by Sven Beer.

that powers energy transfer and primary electron transfer (which in turn drives the whole photosynthetic process) on the one hand, and fluorescence on the other. (See Box 5.3 for a further description of fluorescence.) If so, and if we ignore heat dissipation from the red-photon excitation level as a minor constituent of the de-excitation process, it follows that fluorescence and photosynthesis at large 'compete' for the energy released during de-excitation of chlorophyll. (While the former of course comprises the essential basis for photosynthesis in plants, fluorescence has no direct effect on the process and can be seen as a by-product of de-excitation.) In other words, there is an inverse (or negative) correlation between fluorescence yield (i.e. the amount of fluorescence generated per photons absorbed by chlorophyll) and photosynthetic yield (i.e. the amount of

photosynthesis performed per photons similarly absorbed). This negative correlation can be used as a very important tool for measuring photosynthetic quantum yields as well as rates of photosynthesis (see Section 8.3).

Energy transfer refers to the transfer of excitation energy from one photosynthetic-pigment molecule to another within what we will here call a **photosynthetic antenna**[1] until

---

[1]A photosynthetic unit has classically been described as being made up of PSII and PSI 'antennae', containing the pigment molecules that absorb photons and funnel their energy to neighbouring pigment molecules, and reaction centres of the photosystems that absorb the energy funnelled through the antennae towards them (and the electron carriers of the Z-scheme connecting between the two photosystems). It is debatable, however, how such antennae complexes are connected to one another within the thylakoid membranes. One theory has it that they

Box 5.2   Quantum yields

Quantum yield[2], or what is sometimes designated Φ or, why not simply, **Y** (such as we will often call it in Chapter 8), can in simple terms be defined as the amount of photosynthesis performed per quantum of light (i.e. photon) absorbed by the photosynthetic pigments. Since photosynthesis can be measured in a number of different ways, the integer of Y will vary accordingly. For electron transport in the Z-scheme (see below), 2 photons are needed for each electron (one in photosystem II, PSII, and one in photosystem I, PSI). Assuming 100% efficiency, Y would thus be 0.5. If photosynthesis is measured as O_2 evolution or CO_2 uptake, then it has to be taken into account that each O_2 stems from 2 molecules of H_2O, i.e. 4 electrons have to be transported through the Z-scheme for each O_2 evolved and also for each CO_2 fixed in the Calvin cycle, see, e.g. Figures 5.4 and 6.7), which would require 8 photons altogether. Again, assuming a 100% efficiency, Y would then be 0.125 (1/8). This would be the maximum Y, but since the photosynthetic process is not 100% efficient, the actual value obtained experimentally is around 0.1 (i.e. some 10 photons are required in order for 1 O_2 to be evolved or 1 CO_2 to be fixed and reduced) and use of electrons for other processes such as N assimilation in chloroplasts can drop Y even further. Similarly, the Y for electron transport through PSII (which is commonly measured by PAM fluorometry, see Section 8.3) is maximally 1, but the highest Y for this process ever measured is 0.84. (It is practical to measure and express quantum yield on the basis of electron movement through PSII since a) it gives an estimate of the fraction of energy transferred from photons towards photosynthesis (i.e. 0.84, or 84%, of the photon energy absorbed can maximally be converted to carbohydrates via photosynthesis) and b) this is what is measured by pulse-amplitude modulated (PAM) fluorometry (see Section 8.3). Interestingly, Y is approximately the same for photons of different wavelengths as long as they are within the PAR limits (400–700 nm). At a first glance this may seem strange since, e.g. less of the green photons than of other photons are absorbed in green plants. However, we must then remember that Y is expressed per quanta **absorbed** by the photosynthetic pigments (even if they are few, and Fig. 4.2 shows that **some** photons of green **are** absorbed by chlorophyll), and, again, it is on the basis of those photons that Y is expressed. The fact that high- and low-energy photons yield virtually the same Y was discussed in the above text and is illustrated in Figure 5.1.

[2]There is a distinction between primary quantum yield and overall quantum yield. **Primary quantum yield** refers to the amount of the very first products resulting from photochemical reactions (associated with the primary electron transfer, or charge separation, in the reaction centres, see below) per photon absorbed. In the present text, however, we often extend photosynthetic quantum yield to subsequent processes, and then term it the **overall quantum yield**. Examples of the latter are, e.g. the number of electrons transferred in the Z-scheme (see below), or the amount of O_2 evolved or CO_2 fixed, per photons absorbed by the photosynthetic pigments.

are, indeed, separate units, like separated "puddles", while another has it that these puddles are more or less connected, the extreme being that energy can be freely transferred between the different units analogous to them floating around in a "lake", while the truth for most plants probably lies somewhere in between. The puddle *vs.* lake models are nicely described by Robert Blankenship in his excellent book Molecular Mechanisms of Photosynthesis.

the energy reaches the **reaction centre** where **primary electron transfer** (causing charge separation) starts a flow of electrons from water (the ultimate electron donor) towards CO_2 (the ultimate acceptor) (see Figure 5.2). So, now we need to define the photosynthetic antennae and reaction centres: A photosynthetic

Box 5.3   Fluorescence

Fluorescence in general is the generation of light (emission of photons) from the energy released during de-excitation of matter previously excited by electromagnetic energy. In photosynthesis, fluorescence occurs as electrons of chlorophyll undergo de-excitation, i.e. return to the original orbital from which they were knocked out by photons. Since there is energy loss to processes other than fluorescence (i.e. heat and, mainly, photosynthesis), it follows that the photons of the fluoresced light emitted as the electrons de-excite is of a lower energy level (i.e. higher wavelength) than the photons causing the initial excitation. For chlorophyll, the fluorescence is at ~700 nm whether blue (400–450 nm) or red (650–690 nm) photons caused the excitation. Since fluorescence comprises a small amount of the de-excited energy from chlorophyll, it can usually not be seen by the naked eye[3]. However, it can be 'seen' by fluorometers, which are sensitive instruments that can easily record chlorophyll fluorescence. Since there exists an inverse relationship between photosynthetic yield and fluorescence yield (i.e. how much photosynthesis or fluorescence, respectively, are generated by the photons absorbed by photosynthetic pigments such as chlorophyll), the latter can be used in order to determine the former, and from this photosynthetic rates can be calculated as described in Section 8.3.

[3]Some 80 years ago, the two German scientists H. Kautsky and A. Hirsch **did** notice chlorophyll fluorescence with their eyes as a red colour emitted by various leaves that were illuminated with UV or blue light. This observation is, however, usually not noticeable under normal conditions of irradiance.

antenna (as part of a 'photosynthetic unit') is comprised of up to several hundred pigment molecules (associated with, or embedded in, proteins) that surround a reaction centre, all arranged within the thylakoid membrane such that the energy absorbed from photons is funnelled towards the reaction centre. This energy transfer is called **resonance transfer**, and is non-radiative, i.e. there is almost no energy lost as heat or otherwise in the process. Resonance transfer between molecules can be seen analogously to the first blow of a billiard ball into the triangle of several other balls touching one another; after impact on one ball only, the others spread out on the table after having received its energy. Thus, energy is funnelled throughout the antenna with its various pigment molecules till it reaches the reaction centre at its core. This funnelling through resonance transfer is always directed from pigments that absorb at shorter wavelengths to those that absorb at longer wavelengths in the red or close-to-red band, i.e. from higher to lower energies. The decreasing energy states of the excited pigments towards the reaction centres ensure that no back-flow of energy towards the peripheries of the antennae occurs. A typical cascade of energy flow between pigments in an antenna is carotenoids > chlorophyll b > chlorophyll a > reaction centre.

While energy is bounced from one pigment molecule to the next by resonance transfer, **primary electron transfer** (involving charge separation) occurs only in the reaction centres. There, special chlorophyll a molecules (occurring in pairs) possess the property of not only having their electrons transferred to a higher energy level (or a more peripheral orbit) when excited, but to transfer them out of the

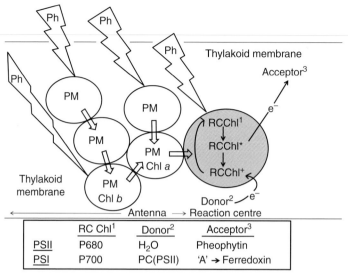

	RC Chl[1]	Donor[2]	Acceptor[3]
PSII	P680	H_2O	Pheophytin
PSI	P700	PC(PSII)	'A' → Ferredoxin

Figure 5.2   Energy transfer and primary electron transfer in the photosystems. As photons (Ph) are absorbed by photosynthetic pigment molecules (PM) of the antenna in the thylakoid membrane (open circles on the left), the energy released by de-excitation of their excited electrons is funnelled, via resonance transfer, through other pigment molecules (open arrows) towards the reaction centre (filled circle on the right). (Note that all pigment chlorophylls of the antenna, as well as the reaction centre, can also receive energy directly by photon excitation.) As the reaction centre chlorophyll (RC Chl) becomes excited (RC Chl*), its electrons are transferred to a primary electron acceptor and the RC Chl becomes positively charged (RC Chl+), a process also called charge separation. When in this state, RC Chl becomes an oxidising agent that can draw electrons from a primary electron donor. The RC Chls, electron donors and electron acceptors of photosystem II (PSII) and photosystem I (PSI) are named in the box on the bottom part of the figure. Drawing by Sven Beer.

molecule to an **electron acceptor**, thus reducing the latter (see Figure 5.2). When doing so, those reaction centres' chlorophylls become positively charged ions (as the electron is released, thus the term charge separation) and act as oxidants as they regain electrons from an **electron donor**. While the reaction centre chlorophyll molecules can also be excited by photons, the previous paragraph indicates that some, and in fact most, of their energy for photochemistry stems from resonance transfer from other pigment molecules 'uphill' in the antenna, with other chlorophyll *a* molecules being in their closest vicinity. Since only photochemistry drives photosynthetic electron flow from H_2O towards the formation of NADPH (see below), one may ask why are not all pho-

tosynthetic pigment molecules of the kind that can perform electron transfer? The answer has to do with the fact that a pigment molecule such as the chlorophyll that constitutes the reaction centre can absorb only a few photons s^{-1} (photons per second) at the photon flux of full sunlight, but because the photochemical reactions there are so fast (some 10 picoseconds, ps, $= 10^{-12}$ s) it makes sense to funnel more photon energy to the reaction centres through resonance transfer than could be supplied to them by direct photon excitation. Also, and perhaps more convincingly, because the chlorophyll *a* molecules of the reaction centres do not readily absorb a large portion of the solar spectrum (i.e. the 450–640 band, cf. Figure 4.2), the involvement of other photosynthetic pigments in light harvesting makes the efficient utilisation of those wavelengths possible.

In some cases, more photon energy is received by a plant than can be used for photosynthesis, and this can lead to photo-inhibition or photo-damage (see later). Therefore, many plants exposed to high irradiances possess ways of dissipating such excess light energy, the most well known of which is the **xanthophyll cycle**. In principle, energy is shuttled between various carotenoids collectively called xanthophylls and is, in the process, dissipated as heat. More details on the xanthophyll cycle as an energy-dissipating mechanism in marine macrophytes are presented in Section 9.1.

In **summary** so far, de-excitation from the red-photon excitation level of the previously photon-excited pigment molecules (Figure 5.1) gives rise mainly to resonance transfer of the released energy to other pigment molecules (some 95% of which eventually reaches the reaction centres) and some fluorescence ($\sim$5%), while very little is dissipated as heat. As the energy reaches the reaction centres at the core of the antenna, the energy causes specialised chlorophyll molecules to transfer electrons to an acceptor molecule (Figure 5.2), thus initiating photosynthetic electron transport from water (see the next section) towards CO_2 reduction (see Section 6.1). In some cases under high irradiances, excess light energy can be dissipated also through other means (e.g. the above-mentioned xanthophyll cycle, see Section 9.1).

It is now time to identify the reaction centre molecules as well as their electron donors and acceptors. There are two types of such reaction centres, one associated with **photosystem II** (**PSII**) and fed by energy funnelled through the antennae (or light-harvesting) complexes of that photosystem, and the other with **photosystem I** (**PSI**) fed by its own antennae. In PSII, the reaction centre consists of a 'special pair' of chlorophyll a molecules called **P680** (the name comes from their absorption peak at 680 nm). In PSI, the corresponding special pair of chlorophyll a is called **P700** (because they absorb maximally at 700 nm). These pigments can either be excited by photons or, as is the much more common case, receive their excitation energy from the surrounding pigment molecules of the antennae (which closest to the reaction centre are other chlorophyll a molecules) via resonance transfer. When this happens, electrons are knocked out of P680 and P700 so as to form **P680$^+$** and **P700$^+$**, respectively (again, a photochemical process termed primary electron transfer or charge separation), and the electrons are transferred to an electron acceptor. As indicated in Figure 5.2, the primary acceptor of electrons released by the reaction centre of PSII is **pheophytin**. Since this compound has a higher (more negative) redox potential than H_2O, it follows that only the excited P680 (often designated **P680***) molecule can reduce it. The ultimate electron donor to P680 is H_2O. This means that as an electron is knocked out of P680, the remaining ion P680$^+$ becomes an oxidising agent that can be re-reduced to P680 by drawing an electron from water. (Again, such electron transfer occurs in pairs of both PSII and PSI reaction centre molecules.) Since H_2O has a very low (very positive) redox potential, it follows that P680$^+$ must be a very **strong oxidant** in order to draw electrons from it; in fact, P680$^+$ is one of the strongest oxidising agents known. In PSI, **plastocyanin** is the electron donor to P700 and an as yet unidentified compound, 'A', is the primary electron acceptor in this photosystem and is reduced by the energised P700 (P700*). 'A' in turn (possibly via some additional electron carriers) reduces a commonly known compound, **ferredoxin**. As electrons are knocked out of P700 to reduce ferredoxin via 'A', the resulting P700$^+$ now becomes an oxidising agent; not as strong an oxidant as P680$^+$, but enough to draw electrons from plastocyanin, the reduced form of which can be seen as the 'end product' of PSII. Thus, and in **summary** so far, the role of the two reaction centres is to boost electrons from H_2O to ferredoxin, the latter then being able to bring about the reduction of NADP+ to form NADPH, which in turn is used to reduce CO_2 to carbohydrates in the CO_2 reduction part of photosynthesis as described in Section 6.1. This electron transfer is done in two steps (i.e. in the two photosystems) because the energy level of the reduced form of the primary electron acceptor of PSII (pheophytin) does not have enough reducing power to reduce ferredoxin. Thus, the tandem action of PSII and PSI are needed to boost the electrons from H_2O to ferredoxin, the latter of which in turn eventually can reduce CO_2. (Some photosynthetic bacteria different from cyanobacteria can make do with PSI only, but they are outside the scope of this book.) More details of the electron transport chain through PSII and PSI are detailed in the following section.

While both excitation, de-excitation, resonance transfer within the antennae and primary electron transfer at the reaction centres are very fast reactions (nano- (10^{-9}) to pico- (10^{-12}) seconds), the very high efficiency of transferring photon energy to cause the oxidation of the electron acceptors in PSII and PSI lies in the arrangement of the pigment molecules and their associated proteins within the thylakoid membrane: The proximity of an energy or electron donor to an acceptor there is not more than 10 Ångström (Å, $= 10^{-10}$ m), which provides for almost all of the energy to be transferred.

5.2   Electron transport

In the photochemical reactions of PSII, the electrons emanating from water reduce P680+ formed by charge separation as the excited P680* reduces pheophytin (Figure 5.2). Thus, as long as there is light, there is a flow of electrons from water through P680 to pheophytin as driven by photon absorption. Before considering the electron flow from pheophytin and onwards, we should add a few words about the way the electrons move from water to P680+: As H_2O is oxidised by P680+, the electrons released travel through a few other compounds on their way to re-reduce P680+. Notably, complexes containing various forms of manganese (Mn) ions are involved in the so-called **water-splitting** or water-oxidising reactions, as are chloride (Cl^-) and calcium ions (Ca^{2+}); interestingly, this is one of the few plant reactions in which Cl^- is essential, which stands in stark contrast to its very high concentration in seawater (see Section 3.3). Thus, it can be viewed that H_2O is 'split' into electrons, protons (H^+) and **molecular oxygen (O_2)** with a simple stoichiometry of 2 molecules of water releasing 4 electrons (e^-), $4 H^+$ and one O_2 (cf. Figure 5.4). (This is the O_2 that is formed as a by-product in photosynthesis, and while it has no positive role in the photosynthetic process, its production can be used by scientists both as a means to measure photosynthetic rates, see Section 8.1, and for respiration, including our breathing.)

On their way through **PSII**, electrons reduce its primary electron acceptor pheophytin. Reduced pheophytin now becomes a reductant that can reduce other compounds on the electrons' way 'downhill' (energy, or redox-potential wise) towards PSI. Within the chlorophyll-protein complexes of PSII, a **quinone (Q_A)** is reduced by pheophytin before transferring the electrons, via another quinone (Q_B) to a third one, **plastoquinone (PQ)**, which is an electron carrier that is fluid within the thylakoid membrane (Figure 5.3). Plastoquinone, in turn, transfers the electrons to another major protein complex of the membrane, the **cytochrome b_6/f complex**,

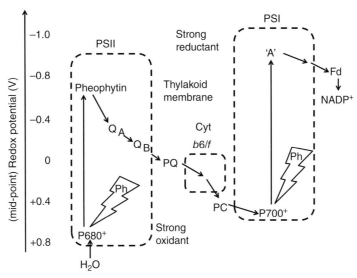

Figure 5.3   The Z-scheme of photosynthetic electron transport. The path of electrons from H_2O to $NADP^+$ via the various carriers of photosystems II (PSII) and I (PSI) is indicated by arrows. The carriers are arranged vertically along a redox-potential y-axis indicating that energy has to be added in the form of photons (Ph) in order to boost electrons 'upwards' from H_2O (via $P680^+$) to pheophytin and from plastocyanin (PC) (via $P700^+$) to 'A'. Note that this figure does not include any stoichiometric aspects (but only the path) of electron flow, nor does it depict the production of protons and O_2 from H_2O; for this, see Figure 5.4; the latter also adds the dimension of spatial arrangements between the different oxido-reductants within the thylakoid membrane (here 'up' and 'down' have no spatial relevance, i.e. the whole scheme could be viewed up-side down). Drawing by Sven Beer.

which transfers the electrons to **plastocyanin**, which is the final electron acceptor of PSII and, thus, the direct electron donor to PSI. When the reaction centres of **PSI** are excited (either directly by photons or by resonance energy transfer through the photosynthetic antenna-pigments of that photosystem), then the resulting $P700^+$ in its reaction centre acts as an oxidant that can draw electrons from plastocyanin. Again, it is the redox potential, location and physical proximity of the various electron-donating and -accepting (reducing and oxidising) carriers within the thylakoid membrane that determine the sequence of electron transport from H_2O to $NADP^+$. Because the electrons can be viewed as 'zigzagging' between compounds of different redox potentials (i.e. from water, through pheophytin, 'down' to plastocyanin and then 'up' to

'A', see Figure 5.3), their path has been termed the **Z-scheme**.

While their exact physicochemical identities are unimportant for the thrust of this book, the characteristics of a few oxido-reductants of photosynthetic electron flow should be described in order to understand subsequent sections: **Plastoquinone** is a diffusible (i.e. not 'fixed', or immobilised) oxido-reductant within the thylakoid membrane. When receiving electrons from Q, it dissociates from PSII and can move towards the cytochrome b_6/f complex, where the electrons are delivered to the latter. When being reduced (by 2 electrons, see Figure 5.4), PQ does not become a negatively charged ion, but draws to it protons (H^+) such that its reduced form is PQH_2; the latter is important for generating an H^+ gradient across the thylakoid membrane (see the following section).

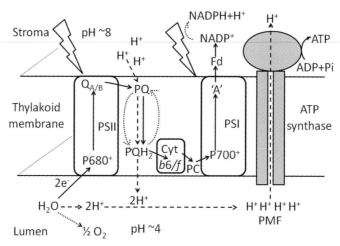

Figure 5.4   The Z-scheme and ATP formation. Electrons derived from H_2O are transported through PSII and PSI just as in the previous Figure 5.3 (full-line arrows). As a result of drawing electrons (e^-) from (or 'splitting') H_2O, protons (H^+) are also released, and they accumulate in the lumen (H^+ movements are indicated by broken-line arrows). Another source of H^+ in the lumen is the so-called 'plastoquinone- (PQ) pump': PQ, as it transports electrons, also transports H^+ from the stroma to the lumen. This generates the proton motive force (PMF) that drives H^+ through ATP synthase whereby ATP is formed. Molecular oxygen (O_2) is formed as a by-product of water splitting. Unlike in Figure 5.3, this figure depicts e^- movements across a thylakoid membrane positioned such that upwards is towards the stroma of the chloroplast and downwards is towards the thylakoid lumen. Drawing by Sven Beer.

Another redox compound worthy of mention is **ferredoxin**. This is an important, almost-terminal, electron carrier of PSI, and when reduced it is a very strong reductant that can reduce $NADP^+$ to NADPH on the stromal side of the thylakoid. (As will be detailed later, ferredoxin can also deliver electrons towards cyclic electron flow around PSI or to O_2 in the Mehler reaction, which is part of what is called the water–water cycle, see Section 5.4.)

The pathway of electrons from H_2O to $NADP^+$ (the Z-scheme) can be envisioned as zigzagging not only across a redox scale (Figure 5.3), but also spatially within the thylakoid membrane (see Figure 5.4). Stoichiometrically, the Z-scheme transfers 4 electrons in order to reduce 2 molecules of $NADP^+$ to NADPH, both of which are needed to reduce 1 CO_2 molecule (see Section 6.1). Since these 4 electrons need to be boosted in two photosystems, 8 photons have to be absorbed in the process (4 for each electron in each photosystem separately), i.e. if we disregard energy losses in the process, then the minimum quantum requirement for reducing 1 CO_2 molecule is 8 (and the quantum yield, Y, is 0.125, see Box 5.1). During the process, 1 O_2 is formed (by the 2 H_2O molecules that released the 4 electrons).

5.3   ATP formation

Not only NADPH is required to reduce CO_2 to carbohydrates, but the chemical energy of ATP is needed as well (as will be detailed in Section 6.1). This ATP is generated in association with photosynthetic electron transport, and therefore depends directly on photon energy; thus, its generation is termed **photophosphorylation**. (In contrast, ATP formation through the NADH-mediated oxidation of sugars and their derivatives by O_2 in mitochondria or within the prokaryotic cells is, accordingly, termed oxidative phosphorylation.) The mechanism by which photophosphorylation generates ATP is similar to that of oxidative phosphorylation in terms of both processes depending on proton gradients across membranes (according to the chemiosmotic hypothesis proposed by Peter Mitchell in the early 1960s). It goes as follows in the case of eukaryotic, chloroplast-containing, plants (see Figure 5.4): Protons accumulate in the lumen of the thylakoids for two main reasons: the **location of the water-oxidising mechanism** and the **properties of plastoquinone (PQ)**. The oxido-reducing compounds that oxidise water (or 'split' water into electrons, H^+ and O_2) are located within the PSII-complex towards the inner side of the thylakoid membrane facing the lumen. Thus, the H^+ formed as water is oxidised accumulates preferably in the aqueous lumen. Plastoquinone, when receiving electrons from Q_B, does so at the outer side of the thylakoid membrane (facing the stroma); this is because Q_B is spatially located to the side of the PSII complex that faces outwards. When 2 electrons reduce PQ, 2 H^+ are also absorbed to the reduced molecule, and these H^+ are thus preferentially taken from the stroma. Conversely, when PQH_2 delivers its electrons to the cytochrome b_6/f complex, i.e. becomes oxidised, the 2 H^+ are released preferentially to the lumen; this is because the site where cytochrome b_6/f receives electrons is located towards the inner side of the thylakoid membrane. In this way, H^+ accumulates within the lumen (is 'pumped' from the stroma to the lumen in a process sometimes called the 'plastoquinone pump') of the thylakoids where they, together with the H^+ from water 'splitting', form what is called a **proton-motive force (PMF)**. The PMF can be seen as a combination of a chemical (because H^+ accumulates in the lumen) and electrically charged (because the protons are positively charged) potential difference between the lumen and the stroma in which the protons strive towards equalising these differences across the thylakoid membrane. In doing so, their only way out happens to be through channels in the fourth large protein complex spanning the thylakoid membranes, i.e. **ATP synthase** (or **ATPase**). (Charged atoms and molecules, i.e. ions, do not easily penetrate lipid bi-layer membranes, see later in Chapter 7, and H^+ is no exception.) The stoichiometry of ATP production is not as straightforward as that of electron transport since H^+ is always around in aqueous solutions. However, there is a need for 3 ATP molecules for each CO_2 reduced in the Calvin cycle, and this is approximately equivalent to 4 H^+ passing through the ATP synthase.

5.4 Alternative pathways of electron flow

The flow of electrons from water through PSII and PSI to $NADP^+$ is often called **linear electron flow**. This is the normal path for electrons, the ultimate purpose of which is to generate NADPH and ATP, which are used to reduce CO_2 to carbohydrates. However, alternative pathways for these electrons also exist. The earliest detected, and most well-known, alternative pathway is that of **cyclic electron flow** around PSI (see Figure 5.5). In this cyclic pathway, electrons from ferredoxin are routed back towards the cytochrome b_6/f complex and onwards to P700 where they are re-excited, and around and around they go. On the way, the reduction of PQ shuttles protons towards the thylakoid lumen such that a proton gradient is formed and ATP is produced (see the previous section). Thus, cyclic electron flow around PSI contributes ATP without producing NADPH. (In some cases, this may also involve what is termed a proton motive 'Q-cycle' in which one electron for the reduction of PQ comes from

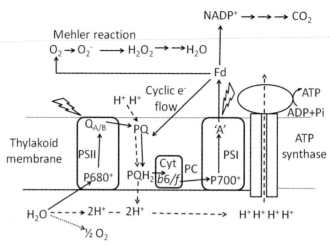

Figure 5.5   The Z-scheme and alternative ways of 'non-linear' electron transport. In addition to linear electron transport (as in Figures 5.3 and 5.4), electrons are here shown to circle around photosystem I (PSI, from ferredoxin, Fd, through the cytochrome b_6/f complex, Cyt b_6/f), as well as reducing O_2 in the Mehler reaction as part of the water–water cycle. See text for details. Full arrows indicate electron movements and broken arrows proton movements. Drawing by Sven Beer.

PSI and the other from the cyclic transfer of electrons involving cytochrome b_6/f.) One possible role of cyclic electron transport is to regulate the ratio of NADPH and ATP formed under conditions where there is extra need for ATP, but the ecological significance of this is still debated. Also, it is likely that this pathway helps to protect the photosystems during limitations of electron flow under, for instance, low-CO_2 conditions within the chloroplasts. It has also been suggested that electrons can **cycle within PSII**. While the latter may protect PSII from photoinhibition at, e.g. high irradiances, the general role of cyclic electron flow is, again, thought to be the addition of ATP under conditions where its production may otherwise limit the photosynthetic reduction of CO_2.

Another path for the electrons is towards reducing O_2, either by a pathway through plastoquinone terminal oxidase (PTOX) or via the more classically known **Mehler reaction** (see Figure 5.5). In the latter, electrons from PSI are routed from ferredoxin, through a 'stromal factor', to O_2, which is reduced to the free oxy-

gen radical superoxide (O_2^-) and then protonated via superoxide dismutase (SOD) to form hydrogen peroxide (H_2O_2); the latter eventually forms water via the activity of ascorbate peroxidase. Since the electrons originally stem from water, and end up in water, the Mehler reaction is part of what is known as the **water–water cycle**. Also here ATP is formed by PQ 'pumping' protons from the stroma to the lumen of the thylakoids (see Section 5.3), but another, perhaps more important, advantage is that free oxygen radicals are 'detoxified': Molecular oxygen (O_2) is potentially harmful if accumulated in cells and tissues, especially since it can form oxygen radicals such as O_2^-; it may therefore be important to reduce the O_2 levels within the cells. High O_2 concentrations that can lead to the formation of oxygen radicals can be formed, e.g. under high-irradiance conditions, especially when CO_2-fixation and -reduction becomes the limiting step in photosynthesis (e.g. for marine plants at high pH values where the CO_2-supply is limiting, see more about such conditions in Part III). A third potential advantage of the water–water cycle is to regulate the proportions of ATP relative to NADPH (in analogy with cyclic electron flow).

In summary of this chapter, the light reactions of photosynthesis use light energy to produce reducing power in the form of NADPH and chemical energy in the form of ATP (the latter compounds being used to reduce CO_2 to carbohydrates, see the next section). This is done by photons first exciting pigment molecules, which after subsequent de-excitation funnel the energy within the antennae to the reaction centres of photosystems II and I (PSII and PSI). There, photochemical reactions cause electrons to move against an energy gradient, and subsequently along an energy gradient (in each photosystem), so generating a flow of electrons from H_2O (the electron donor to the whole process), through electron carriers of the so-called Z-scheme, towards reducing $NADP^+$ to NADPH. Protons (H^+), which are also released as electrons are taken from H_2O, accumulate in the lumen of the thylakoids, and so do H^+ released as the reduced form of the electron carrier plastoquinone (PQH_2) transfers electrons to PSI. These H^+ generate a proton-motive force (PMF) that drives them from the lumen of the thylakoid to the stroma of the chloroplast through a transmembrane ATP synthase, whereby ATP is generated.

The red alga *Delesseria sanguinea* at the Swedish west coast. Photo by Katrin Österlund.

Chapter 6

Photosynthetic CO$_2$-fixation and -reduction

In this chapter we will in Section 6.1 first examine how CO$_2$ is converted into sugars; this is, in all plants, done via the Calvin cycle. Central to the Calvin cycle is the primary CO$_2$-'fixing' (or carboxylating) enzyme ribulose-bisphosphate carboxylase/oxygenase (Rubisco). However, as its name implies, this enzyme not only acts as a carboxylase, but can also act as an oxygenase (i.e. fix O$_2$). When doing so, one of its products is metabolised further in a process called photorespiration, in which some of the CO$_2$ previously reduced to sugars is lost from the plant. Photorespiration is a consequence of today's low concentration of CO$_2$ and high concentration of O$_2$ compared with Rubisco's relative affinity (mirrored in its K_m values) for those two substrates. Terrestrial plants growing in hot and arid environments must, at least partly, close their stomata during the day so as to hinder detrimental water loss. Given Rubisco's qualities, the carboxylation reaction of the Calvin cycle would then be severely hindered were it not for additional CO$_2$-fixing systems developed in these plants called C$_4$ metabolism (in C$_4$ plants) or Crassulacean acid metabolism

(CAM, in CAM plants). These auxiliary systems function as CO$_2$-concentrating mechanisms (CCM), which provide Rubisco with much higher CO$_2$ concentrations than would be possible by diffusion through the (partly closed) stomata only. Such CCMs are described in Section 6.2 (while the CCMs of marine plants are described in Chapter 7).

6.1 The Calvin Cycle

In order to 'fix' CO$_2$ (= incorporate it into organic matter within the cell) and reduce it to sugars, the NADPH and ATP formed in the light reactions are used in a series of chemical reactions that take place in the stroma of the chloroplasts (or, in prokaryotic autotrophs such as cyanobacteria, the cytoplasm of the cells); each reaction is catalysed by its specific enzyme, and the bottleneck for the production of carbohydrates is often considered to be the enzyme involved in its first step, i.e. the fixation of CO$_2$ (see below). These

Photosynthesis in the Marine Environment, First Edition. Sven Beer, Mats Björk and John Beardall.
© 2014 John Wiley & Sons, Ltd. Published 2014 by John Wiley & Sons, Ltd.
Companion Website: www.wiley.com/go/beer/photosynthesis

CO_2-fixation and -reduction reactions are known as the Calvin cycle (named after Melvin Calvin, who together with Andrew Benson and James Bassham discovered them in the late 1950s, see Box 6.1) or the C_3 cycle (see Figure 6.1). The latter name stems from the fact that the first stable product of CO_2 fixation in the cycle is a 3-carbon compound called phosphoglyceric acid (PGA): Carbon dioxide in the stroma is fixed onto a 5-carbon sugar called ribulose-bisphosphate (RuBP) in order to form 2 molecules of PGA (see [1] in Figure 6.1). It should be noted that this reaction does not produce a reduced, energy-rich, carbon compound, but is only the first, 'CO_2-fixing', step of the Calvin cycle. In subsequent steps, PGA is energized by the ATP formed through photophosphorylation and is reduced by NADPH (that was formed in the last step of photosynthetic electron transport) to form a 3-carbon phosphorylated sugar (glyceraldehyde 3-phosphate, also called phosphoglyceraldehyde), here denoted simply as triose phosphate (TP); these reactions can be called the CO_2-reduction step of the Calvin cycle (see [2] in Figure 6.1). At this point, 1/6 of the TPs formed leave the cycle while 5/6 are needed in order to re-form RuBP molecules in what we can call the regeneration part of the cycle (see [3]); it is this recycling of most of the final product of the Calvin cycle (i.e. TP) to re-form RuBP that lends it to be called a biochemical 'cycle' rather than a pathway. It should be noted that while 2 molecules of NADPH and 2 of ATP are needed in order to reduce the 2 PGAs formed by the fixation of 1 CO_2, an extra ATP is needed in order to phosphorylate a mono-phosphorylated sugar in the regeneration phase so as to form the bi-phosphorylated sugar RuBP. Thus, the simplistic stoichiometry

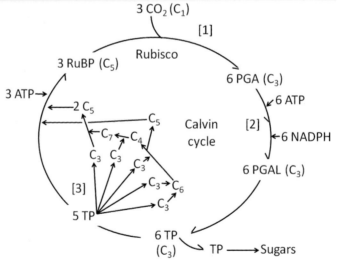

Figure 6.1 **The Calvin (or C_3) cycle of CO_2-fixation and -reduction.** The figure shows the CO_2-fixation stage [1] of carboxylation via ribulose-bisphosphate carboxylase/oxygenase (Rubisco), the reduction stage [2] where ATP and NADPH from the light reactions are used in order to form the reduced sugar phosphoglyceraldehyde (PGAL), which is a triose phosphate (abbreviated as TP), and the regeneration stage [3], in which 3 molecules of RuBP is reformed by 5 TPs. The numbers in front of the compounds indicates their stoichiometric relationships to one another; the subscripts to Cs within brackets denote the number of carbon atoms in each compound. Drawing by Sven Beer.

Box 6.1   Calvin's experiment

In the 1950s, Melvin Calvin and co-workers performed a series of what are termed ^{14}C-pulse experiments at the Berkeley National Laboratories, USA, as follows (see Figure 6.2.): Photosynthesising cells of *Chlorella* (a freshwater unicellular green alga) were kept in an illuminated flattened glass flask (called a 'lollipop' flask) and ^{14}C-containing CO_2 (probably in the form of HCO_3^-) was added during steady-state photosynthesis (see panel a of the figure below). A stopper was then opened after various time intervals and part of the contents of the flask were thus emptied into a beaker containing boiling alcohol; the hot alcohol both killed the cells and extracted their organic compounds. After concentration, the extract was applied as a spot on a corner of large square filter papers, and the organic cellular constituents were separated by two-dimensional paper chromatography (according to their solubility properties in various solutes). Those compounds that contained ^{14}C-label were visualised by placing X-ray-sensitive films on the dried chromatograms (causing dark spots to appear on the films, which are also called autoradiograms, see panel b of the figure), and those compounds were then identified by comparisons with chromatograms containing known standards. After 5 s of [14]CO_2 exposure, phosphoglyceric acid (PGA) contained most of the label, and this was thus determined to be the first compound formed in the fixation of CO_2. As [14]CO_2-exposure times increased, other compounds were identified, and the sequence of the Calvin cycle could thus be determined. It is interesting to note that although photosynthetic research today is carried out mainly on terrestrial plants, some early basic discoveries such as the Calvin cycle were made using a unicellular aquatic alga.

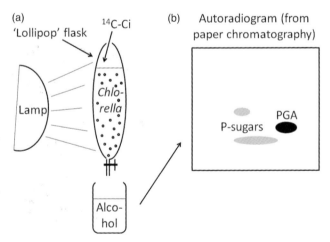

Figure 6.2   Calvin's experiment. Melvin Calvin and co-workers fed ^{14}C-containing inorganic carbon (^{14}C-Ci) to *Chlorella* cells for various amounts of time during steady-state photosynthesis, and the cells were then killed in boiling alcohol (a), where after all organic compounds in the cells were extracted and separated using paper chromatography (b, showing the autoradiogram from the paper chromatogram). The first compounds that contained ^{14}C label were phosphoglyceric acid (PGA) and, soon thereafter, phosphorylated sugars (P-sugars). Drawing by Sven Beer.

of the Calvin cycle calls for 2 NADPH and 3 ATP for each CO_2 fixed and reduced. With 3 CO_2 required for every TP leaving the cycle to form other sugars, 6 NADPH and 9 ATP are required for each such TP (see Figure 6.1).

The regeneration phase of the Calvin cycle, i.e. where RuBP is re-formed from TP, is complex and will not be detailed here. In principle, 5 molecules of TP are rearranged so as to form 3 molecules of RuBP (and ATP is used in the process). The principle of reorganising the 5 TPs is illustrated in the lower left corner of Figure 6.1.

The stoichiometry of CO_2-fixation and -reduction can logically be followed if we start with 3 molecules (or mol, which forms a more logical number of molecules) of CO_2. Then, since 1 CO_2 forms 2 PGA, the result will be 6 PGAs, which in turn form 6 TPs. Out of these 6 TPs, 5 continue to the RuBP-regeneration part of the cycle (see Figure 6.1) while 1 TP is the net production of the cycle which, after leaving the chloroplast, is available for the formation of other sugars. So, for example, 6 turns of the Calvin cycle (fixing 1 CO_2 in each turn) will produce a net of 2 TPs, which in principle can form a 6-carbon sugar such as glucose or fructose, while 12 turns produces enough TPs to give rise to a molecule of sucrose; the latter is indeed considered an end product of photosynthesis (while, again, the end product of the Calvin cycle is TP). Of course, the Calvin cycle keeps turning and producing millions of TPs as long as the light is on so that ATP and NADPH are generated for the reduction of CO_2; when the light is turned off, the cycle will stop after a few seconds as ATP (formed by photophosphorylation) and NADPH (formed by electron flow via the Z-scheme) are depleted within the chloroplasts (or the prokaryotic photosynthesising cells). Finally, it should be realised that the formation of the final carbohydrates within the photosynthesising cell, as well as the re-formation of RuBP, is a complex process with many steps, but those are out of the scope of this

book; again, the interested reader is referred to the same books mentioned in the end of Chapter 4. Under conditions where the rate of photosynthetic production exceeds the demand of the plant, some photosynthate (= product of photosynthesis) can be stored in the cells (as, e.g. starch in green algae and higher plants); this part of the photosynthetic process is also out of the scope of this book.

Since the enzyme that is involved in the first part of the Calvin cycle (i.e. that part in which CO_2 is fixed onto RuBP) is often considered as the bottleneck for photosynthetic production, it requires some consideration here. This enzyme is called ribulose-bisphosphate carboxylase/oxygenase, or **Rubisco**, and the carboxylase part of its name (ribulose-bisphosphate **carboxylase**) shows that it catalyses the fixation of CO_2 onto RuBP (which is a carboxylation reaction). (See Figure 6.3 for an illustration of Rubisco.) **John Raven** in one of his papers (Raven and

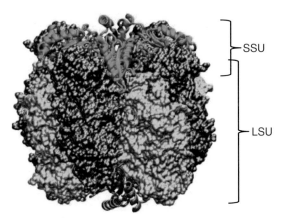

Figure 6.3 A reconstructed model of a **ribulose-bisphosphate carboxylase/oxygenase (Rubisco) molecule,** comprised of 8 large subunits (LSU, 4 of which are seen in yellow and grey) and 8 small subunits (SSU, some of which are seen in green and brown); this form of Rubisco is, accordingly, termed L_8S_8. With permission from, and thanks to, Dr. Fabrice PA David, STB, Swiss Institute of Bioinformatics. Data from the Protein Data Bank (1RCO).

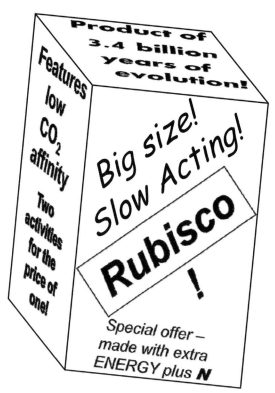

Figure 6.4   A cereal box model of Rubisco. Drawn after an idea of John Raven (with his kind permission). Drawing by John Beardall.

Johnston, 1991) presented a "one-box model" for Rubisco in which its 'virtues' were extolled as one might market a new brand of cereal (see Figure 6.4), i.e. as a) **big, made with extra energy and N**, b) **slow acting** (in reality low affinity for CO_2), and c) providing **two reactions for the price of one** (in reality an oxygenase as well as carboxylase). We (SB and JB) have often pondered that description in our Marine Botany and Marine Photosynthesis courses as follows: That Rubisco is **big** is a matter of fact. With a molecular weight (MW) of some 560 kiloDalton (kDa), most forms of Rubisco are comprised of 8 large and 8 small subunits (termed L_8S_8). This size represents a considerable investment by cells, both in N and in energy required for protein

synthesis, especially given that relatively large amounts of the enzyme are required to allow the cells to assimilate CO_2 at a rate sufficient to explain their achieved specific growth rates. Interestingly, while the large subunits are encoded for by the chloroplastic genome (and they are accordingly transcribed and translated within the chloroplasts of eukaryotic plants; remember that chloroplasts have their own genome(?); if not, see the description of endosymbiosis in Chapter 1), the small subunits are encoded for by the nuclear genome. The latter are then transported into the chloroplasts, where the whole, large, enzyme (or holoenzyme) is assembled. Incidentally, "most forms of Rubisco..." is an overstatement since this enzyme is one of the most conserved ones we know. It is thought that the cyanobacterial Rubisco was formed more or less as it is today with the first prokaryotic organisms billions of years ago, and the Rubisco of higher plants is very similar to the one found in those cyanobacteria even though its kinetic properties have evolved somewhat (e.g. the $K_m(CO_2)$ of some 'recent' terrestrial plants can be as low as 10 µM).

The second of John Raven's attributes to Rubisco, i.e. that it is slow acting, is substantiated by its relatively **low affinity for CO_2**. The 'best' Rubisco we know has a $K_m(CO_2)$ of ~10 µM. If we look at this K_m as the concentration of CO_2 where the catalysed reaction speed is half its maximal one, then this means that today's approximate atmospheric CO_2 concentration, when dissolved in an air-equilibrated photosynthesising cell (~13 µM, see Section 3.2), only about half-saturates the carboxylation reaction (see Figure 6.5). Rubiscos of other photosynthesising organisms have an even lower affinity for CO_2 (higher $K_m(CO_2)$): Those species that exhibit a CO_2-concentrating mechanism (CCM, see the next section), including most marine plants and cyanobacteria (see Chapter 7), can have Rubiscos

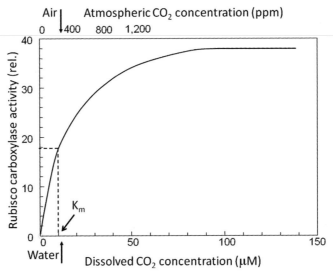

Figure 6.5 **The carboxylase activity of Rubisco** having a $K_m(CO_2)$ of 10 μM (where the reaction rate is half the maximal one, marked on the lower x-axis), as a function of dissolved (in the chloroplast stroma if the photosynthesising cell is in equilibrium with the atmosphere) CO_2 concentrations (lower x-axis). The corresponding aerial CO_2 concentration is given on the upper x-axis. Today's atmospheric (394 ppm) and water-equilibrated (at 20 °C, 13 μM) CO_2 concentrations are indicated by the down- ('Air') and up-ward ('Water') arrows, respectively. Drawing by Sven Beer.

with $K_m(CO_2)$ values of 50 μM or even 200–300 μM (for cyanobacteria). This means that without CCMs, the present-day CO_2 concentration is so low that those organisms would hardly be able to photosynthesise at all.

The third characteristic attributed to Rubisco is that it provides '2 for the price of one'. For Rubisco, this means that the enzyme not only functions as a carboxylase, but that it also **acts as an oxygenase** (the last 'o' in Rubisco designates its oxygenase activity).

The selectivity factor S_{rel} defines the relative rates of carboxylase and oxygenase reactions catalysed by Rubisco and is given by the following equation:

$$S_{rel} = \frac{K_{1/2}(O_2)k_{cat}(CO_2)}{K_{1/2}(CO_2) \cdot k_{cat}(O_2)} \quad (6.1)$$

where $k_{cat}(CO_2)$ is the CO_2-saturated specific rate of carboxylase activity of Rubisco (mol CO_2 mol^{-1} active sites), $K_m(CO_2)$ the concentration of CO_2 at which the CO_2 fixation rate by Rubisco is half of $k_{cat}(CO_2)$, $k_{cat}(O_2)$ the O_2-saturated specific rate of oxygenase activity of Rubisco (mol O_2 mol^{-1} active sites) and $K_m(O_2)$ is the concentration of O_2 at which the O_2 fixation rate by Rubisco is half of $k_{cat}(O_2)$.

When Rubisco reacts with oxygen instead of CO_2, instead of 2 molecules of the 3-carbon compound PGA being formed (see Figure 6.1), only 1 molecule of PGA is formed together with 1 molecule of the 2-carbon compound **phosphoglycolate** as 1 molecule of O_2 is fixed to the 5-carbon compound RuBP (see Figure 6.6). Not only is there no gain in organic carbon by this reaction, but CO_2 is actually lost in the further metabolism of phosphoglycolate, which comprises a series of reactions termed **photorespiration** (Figure 6.6). The respiration part of the word stems from the fact that, like in mitochondrial, or dark, respiration, O_2 is

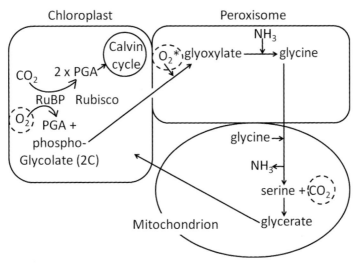

Figure 6.6 **The pathway of photorespiration in higher plants.** As O_2 (rather than CO_2) is fixed via ribulose-bisphosphate carboxylase/oxygenase (Rubisco), one of the products, the 2-carbon (2C) compound phospho-glycolate, is metabolised further in peroxisomes and mitochondria. In all, O_2 is consumed and CO_2 evolved in the process (hatched circles), and since the substrate ribulose-bisphosphate (RuBP) is generated in photosynthesis by the Calvin cycle, the process is termed photorespiration. Note that O_2 may (in the glycolate oxidase variant of the cycle, marked with *) or may not (in the glycolate dehydrogenase type) be consumed also in the peroxisome. Drawing by Sven Beer.

consumed and CO_2 is produced. However, unlike mitochondrial respiration, RuBP is one of the substrates in photorespiration (with O_2 being the other), and since this compound is generated by the Calvin cycle in the light (see above), this has been termed photorespiration. The photorespiratory cycle can take two forms as far as the metabolism of phosphoglycolate is concerned: the latter compound can either be oxidised via glycolate oxidase or metabolised via glycolate dehydrogenase.

While photorespiration is a complex process, involving reactions occurring also in cellular compartments other than the chloroplast (see Figure 6.6), it is also an apparently wasteful one in which a product of photosynthetic CO_2-fixation and -reduction (RuBP) is oxidised back to CO_2, and it is not known why this process has evolved in plants altogether. While most marine algae in effect do not feature photorespiration (although its biochem-

ical pathway usually is present), it appears that some seagrasses do (see chapter 7d). Photorespiration can reduce the net photosynthetic production by up to 25%. When discussing this topic, however, one must remember that when Rubisco evolved, the much higher CO_2 and lower O_2 concentrations (see Chapter 1) caused higher rates of carboxylation *vs.* oxygenation such that photorespiration became much less apparent. So, while today's atmosphere of low CO_2 and high O_2 concentrations causes losses to the photosynthetic productivity of most terrestrial plants due to Rubisco's oxygenase activity, the enzyme may still be (or apparently is) the best alternative for photosynthetic production via the Calvin cycle. There is indeed a view now that perhaps Rubisco is better attuned to environmental pressures than had previously been thought possible and that selective pressure over the course of evolution has resulted in at least a partial adaptation of Rubiscos in different groups of photoautotrophs to differing CO_2/O_2 conditions. Thus, Rubisco evolved when CO_2 levels were very much higher than today, and this is reflected in the high $K_m(CO_2)$ values in cyanobacteria and chemolithotrophic bacteria, but as CO_2 levels in the atmosphere have decreased over geological time, more recently evolved species show adaptations in $K_m(CO_2)$, with values as low as 10 μM.

The energy costs of phosphoglycolate synthesis and metabolism are fairly straightforward to estimate. The 2 C in each phosphoglycolate are derived from 2 of the 5 carbons in the RuBP that reacts with oxygen. Thus, to synthesise one phosphoglycolate molecule requires the energy needed to make 2/5 of the RuBP molecule plus the costs of making the other 3/5 of the RuBP that are used to form TP. Fixation of 2 CO_2 via the Calvin cycle, to form the phosphoglycolate itself, uses 4 NADPH and 6 ATP and the reduction of one PGA molecule to TP costs an extra NADPH and one ATP. Thus, the cost per molecule of phosphoglycolate made is 5 NADPH and 7 ATP. This is a basic cost, independent of the pathway used to further metabolise phosphoglycolate. If one phosphoglycolate is metabolised via glycolate oxidase (in seagrasses and the brown and red algae) to half a molecule of TP and half a molecule of CO_2, then this incurs an energy cost of one NADPH and 1.5 ATP. In the case of glycolate dehydrogenase, this cost is one NADPH. Adding up the NADPH and ATP costs for producing the half molecule of TP, which can be produced from each molecule of phosphoglycolate we get 6 NADPH and 8.5 ATP molecules for the phosphoglycolate oxidase variant of the photorespiratory cycle and 6 NADPH and 7 ATP molecules for the glycolate dehydrogenase variant. Cyanobacteria are believed to not have a complete photorespiratory cycle, but given the ubiquitous and high level of expression of CCM activity (see Chap-

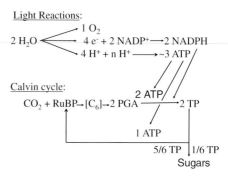

Figure 6.7   The stoichiometry of photosynthesis. See the above text for explanations. Drawing by Sven Beer.

ter 7), which would suppress oxygenase activity in these species anyway, we will not consider the energy costs of photorespiration in cyanobacteria.

In summary of Chapters 5 and 6 till now, and as a way of recapitulating the main flow of energy, CO_2 and important intermediates, we can outline the stoichiometry of photosynthesis as follows (see Figure 6.7): If we (or the plant, or just nature) 'wish(es)' to reduce 1 molecule (or mol; these units will be used alternatively, or will largely be skipped altogether, in the following paragraph) of CO_2, then we start the light reactions with 2 molecules of water. These, when 'split' in PSII, yield 4 electrons, 4 H^+ and 1 O_2. In linear electron transport, the electrons are transported through the two photosystems, after which they reduce 2 molecules of $NADP^+$ to NADPH (together with 2 H^+, but we don't count those so exactly here[1]). The H^+ released from 1 H_2O, together with H^+ from other sources (e.g. from the PQ 'pump', and other protons around, together marked as

[1] As a student, I (SB) never understood why protons (H^+) are never included in stoichiometric relationships (and I still don't). The answer I usually get when asking is that H^+ are "always around", which of course is true for every liquid (a glass of water contains many protons), and so "we don't count them". Well, so be it until I get a better explanation.

n H^+) that generate the PMF in the lumen of the thylakoids will cause ~3 ATP molecules to be formed as they pass through the ATP synthase (this is assumed rather than measured since there is a need for 3 ATPs for each 'turn' of the Calvin cycle). So, the light reactions will generate 2 NADPH, ~3 ATP and, as a by-product, 1 O_2. The 2 NADPs and 2 of the ATPs are then used for reducing 1 molecule of CO_2 in the Calvin cycle (by carboxylation of 1 molecule of RuBP, a 5-carbon compound) via Rubisco to generate 2 molecules of TP. 1/6 of the TPs formed will leave the Calvin cycle as its final product (and then form other sugars), while 5/6 of the TPs will be used for regenerating RuBP. For this regeneration process, one extra ATP is used for each RuBP formed. If so, then the stoichiometry of photosynthesis is **2H_2O / 4e^- / 1O_2 / 2NADPH / 1CO_2 / 3ATP / 8 photons** (the 8 photons are needed if we assume a 100% efficiency in their ability to move electrons from water to $NADP^+$: 4 photons for the 4 electrons through PSII and 4 more photons to move those same electrons through PSI; these photons are not marked in Figure 6.7). Note that protons are not included in stoichiometric relationships such as the one described here. In order to arrive at the stoichiometry depicted in Figure 6.1, we have to reduce 3 molecules of CO_2, which would require 6 molecules of H_2O etc.

6.2 CO_2-concentrating mechanisms

Because of Rubisco's low affinity to CO_2 as compared with the low atmospheric, and even lower intracellular, CO_2 concentration (see Table 7.1), systems have evolved in some plants by which CO_2 can be concentrated at the vicinity of this enzyme; these systems are accordingly termed **CO_2 concentrating mechanisms**

(**CCM**). For **terrestrial plants**, this need for concentrating CO_2 is exacerbated in those that grow in hot and/or arid areas where water needs to be saved by partly or fully closing stomata during the day, thus restricting also the influx of CO_2 from an already CO_2-limiting atmosphere. Two such CCMs exist in terrestrial plants: the **C_4 cycle** and the **Crassulacean acid metabolism** (**CAM**) **pathway**. Since these CCMs were discovered and studied before those of aquatic plants, we will describe them in this section; the CCMs of marine plants, which are more relevant for this book, and are very different from the terrestrial mechanisms, will be discussed in Chapter 7.

The **C_4 cycle** is called so because the first stable product of CO_2-fixation is not the 3-carbon compound PGA (as in the Calvin cycle) but, rather, malic acid (often referred to by its anion malate) or aspartic acid (or its anion aspartate), both of which are 4-carbon compounds. In this biochemical cycle (see Figure 6.8), CO_2 enters the **mesophyll cells** (which are located just beneath the epidermis that contains the stomata, i.e. close to the atmosphere) and is fixed onto the 3-carbon molecule phosphoenolpyruvate (PEP) by the enzyme **phosphoenol-pyruvate carboxylase** (**PEP-case**). While the first compound formed by this carboxylation is oxaloacetic acid (or its anion oxaloacetate), which is also a 4-carbon compound, this is immediately reduced to malate via the enzyme malate dehydrogenase or is, in another step, aminated to form aspartate (depending on the species of C_4 plant). The 4-carbon compounds malate or aspartate are then transported to another type of cells, the **bundle-sheath cells** (which are located more in the middle of a cross section of the leaf, i.e. they are more isolated from the atmosphere) where they are **decarboxylated** to release the CO_2 that was originally fixed via PEP-case in the mesophyll cells; this CO_2 is then **re-fixed and reduced via the Calvin**

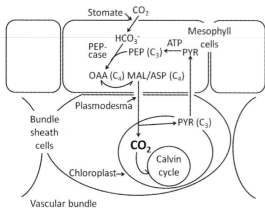

Figure 6.8 **C$_4$ metabolism in terrestrial C$_4$ plants.** CO$_2$, in the form of HCO$_3^-$, is 'carboxylated' onto the 3-carbon (C$_3$) compound phosphoenopuryvate (PEP), as catalysed by phosphoenolpyruvate carboxylase (PEP-case), to form the 4-carbon compound oxaloacetic acid (OAA), the latter of which is transformed either to malic acid (or its anion malate, MAL) or, via an additional transamination, to the amino acid aspartic acid (or aspartate, ASP) (depending on plant species), in the mesophyll cells. The latter two compounds are then transported to the bundle sheath cells through plasmodesmata (openings between adjacent cells), where they are 'decarboxylated' to form high concentrations of CO$_2$, which is further metabolised in the Calvin cycle of their chloroplasts. Pyruvate (PYR), the other product of the decarboxylation reaction, is transported back to the mesophyll cells, where it is phosphorylated to form PEP, thus closing the cyclic event of this metabolism (also called the C$_4$ cycle). Drawing by Sven Beer.

cycle in the chloroplasts of the bundle-sheath cells. At the same time that CO$_2$ is released, the decarboxylation reaction also generates pyruvate, which is transported back to the mesophyll cells where it is phosphorylated back to PEP (with the use of ATP), which can then again serve as the substrate for fixing atmospheric CO$_2$; thus C$_4$ metabolism acts in a cyclic manner (and is therefore called the **C$_4$ cycle**) that reforms the substrate for CO$_2$ fixation (just like the Calvin cycle reforms its substrate RuBP).

In order to understand how the C$_4$ cycle can act as a CO$_2$-**concentrating** mechanism, some of its characteristics need to be pointed out: 1) The Ci form that is actually fixed to PEP via PEP-case is HCO$_3^-$; this is the major ionic form of Ci at pH 6–9 (see Section 3.2), which includes the likely pH in the mesophyll cells. Assuming an intracellular pH of 8, and a fast, possibly carbonic anhydrase (CA)-mediated equilibration between dissolved CO$_2$ and HCO$_3^-$, the concentration of the latter ion is some 100 fold higher than that of CO$_2$. Thus, instead of the ca. 13 µM dissolved CO$_2$ that is in equilibrium with the air[2], carboxylation via PEP-case provides the carboxylating enzyme with a much higher concentration of Ci; 2) PEP-case has thus a high affinity for Ci with a K_m(HCO$_3^-$) of 0.2–0.4 mM, equivalent to around 4-8 µM CO$_2$ at pH 8, and the level of PEP-case in C$_4$ plants is usually so high as to not be rate limiting; 3) Unlike Rubisco, PEP-case features no oxygenase function. Thus, O$_2$ does not compete or otherwise interact with atmospheric CO$_2$ in this type of carboxylation; 4) The compartmentalisation of atmosperic-CO$_2$ fixation (in the mesophyll cells, which are located close to the atmosphere) and the very efficient decarboxylation, which re-forms CO$_2$ (in the bundle-sheath cells, which are more isolated from the atmosphere), allows for a high concentration of CO$_2$ to surround Rubisco in the chloroplasts of the latter cells. Thus, Rubisco becomes saturated (or near-saturated) with CO$_2$, and O$_2$ can hardly compete with CO$_2$ at the catalytic

[2]While ca. 13 µM is the concentration of CO$_2$ formed in equilibrium with today's atmosphere, its intracellular concentration is likely much lower as limited by its inward diffusion through the stomata openings as CO$_2$ is consumed by photosynthesis. This limitation is exacerbated by the stomata often being partly closed so as to save water, and so the true intracellular CO$_2$ concentration is in effect substantially less than 13 µM (but then there is the CCM that concentrates CO$_2$ to Rubisco in the Calvin cycle!).

site of the enzyme, thus eliminating photorespiratory losses of CO_2. Under such circumstances, Rubisco in C_4 plants can perform carboxylation at higher rates than in C_3 plants, even at low influx rates of CO_2 through partly closed stomata during, e.g. water-stress conditions in the arid areas where these plants usually grow. C_4 plants are, indeed, more common in areas of high temperature, especially when accompanied with scarce rains, than in areas with higher rainfall; typical C_4 plants include many arid-area grasses, including the economically important sugarcane and maize (or corn in the US).

While atmospheric CO_2 is fixed (in the form of HCO_3^-) via the C_4 cycle, it should be noted that this biochemical cycle cannot reduce CO_2 to high energy containing sugars; rather, organic (malate) or amino (aspartate) acids are formed by this carboxylation. Thus, and since the Calvin cycle is the only biochemical system that can reduce CO_2 to energy-rich carbohydrates in plants, it follows that the CO_2 initially fixed by the C_4 cycle and released in the bundle-sheath cells by the decarboxylation of malate or aspartate is finally reduced via the Calvin cycle also in C_4 plants. In summary, the C_4 cycle can be viewed as being an additional CO_2 sequesterer, or a biochemical CO_2 'pump', that concentrates CO_2 for the rather inefficient enzyme Rubisco in C_4 plants that grow under conditions where the CO_2 supply is extremely limited because partly closed stomata restrict its influx into the photosynthesising cells.

Crassulacean acid metabolism (CAM) is similar to the C_4 cycle in that atmospheric CO_2 (in the form of HCO_3^-) is initially fixed via PEP-case into the 4-carbon compound malate. However, this fixation is carried out during the night when the stomata of these plants are open, and since photosynthesis cannot be performed in the absence of light, the malate formed is stored in the cell vacuoles (see Figure 6.9). During the day, the stored malate is moved from the vacuoles to the cytoplasm, where it is decarboxylated to form CO_2 and pyruvate; the former is then carboxylated via Rubisco and reduced in the Calvin cycle in the chloroplasts. During this time, the stomata are closed[3], so the only CO_2 supply for photosynthesis stems from the malate that was formed from atmospheric CO_2 during the night. As in C_4 plants, the carboxylation of CO_2 into a C_4 compound (here during the night) and its subsequent decarboxylation (during the day) is a mechanism that concentrates CO_2 at the active site of Rubisco. Unlike C_4 plants, CAM plants do not recycle the pyruvate formed by the decarboxylation of malate into PEP; instead, PEP for carbon fixation is derived from the breakdown of starch, and so the biochemical path of CO_2 fixation is here not a 'cycle' but, rather, a (linear) 'pathway'.

The ecological advantage behind CAM metabolism is that a CAM plant can grow, or at least survive, under prolonged (sometimes months) conditions of severe water stress. By opening the stomata during the much cooler night rather than during the hot day, water loss is minimised while CO_2 can still enter the plant cells. However, the vacuoles of CAM plants have a limited storage capacity (for the CO_2 equivalents in malate), and so the malate accumulated therein during the night is usually depleted before noon. Thus, photosynthesis in CAM plants functions only for a limited number of hours during the day, and this may be the reason that these plants feature slow growth in spite of their CCM. CAM plants are typical of the desert flora, and include most cacti. The almost-only crop plant featuring CAM

---

[3]Some CAM plants are facultative, i.e. they do close the stomata during the day under harsh drought conditions, but can leave them at least partially open during conditions of a larger water supply, e.g. during rainy seasons. In the latter conditions, they thus function more or less as regular C_3 plants.

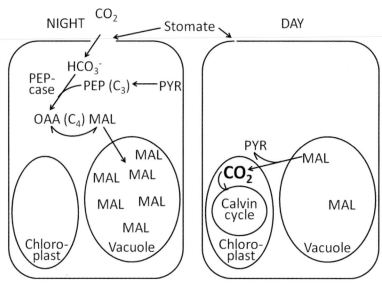

Figure 6.9   CAM metabolism in terrestrial CAM plants. CO_2 enters the photosynthesising cells during night-time, when the stomata are open, and is, like in C_4 plants (see Figure 6.8), 'carboxylated' onto phosphoenolpyruvate (PEP) via phosphoenolpyruvate carboxylase (PEP-case) to form the ions of malic acid malate (MAL). These accumulate in the vacuoles during the night, and are in the morning transported out of the vacuoles, are decarboxylated, and the CO_2 released is metabolised in the light via the Calvin cycle. Unlike in C_4 plants, PEP is not formed by recycling pyruvate (PYR), and so the C_4-like metabolism of CAM plants does not involve a C_4 'cycle' *per se*. Drawing by Sven Beer.

metabolism is pineapple, but the recent use of young *Opuntia*-cacti phylloclades as 'fruits' adds those cacti to this category.

In **summary** of this section, there are ways in which certain plants can concentrate CO_2 to the rather inefficient Rubisco of the Calvin cycle, the latter of which is their only way to reduce CO_2 to high-energy carbohydrates. The two ways of doing so are in terrestrial plants the C_4 cycle and CAM metabolism. While both feature CO_2-fixation via PEP-carboxylase, this reaction does not generate sugars, and so this type of carboxylation is only an aid for enhancing the supply of CO_2 to Rubisco in the Calvin cycle. The principal difference between C_4 and CAM metabolism is that in C_4 plants the initial fixation of atmospheric CO_2 and its final fixation and reduction in the Calvin cycle is separated in space (between mesophyll and bundle-sheath cells) while in CAM plants the two processes are separated in time (between the initial fixation of CO_2 during the night and its re-fixation and reduction during the day).

In a short **summary** of the last two chapters, the light reactions use the energy of photons absorbed by the photosynthetic pigments to form ATP (by photophosphorylation) and NADPH (by electron transport in the Z-scheme). These two compounds are then used for the reduction of CO_2 to sugars in the CO_2-fixation and -reduction reactions known as the Calvin cycle. In some plants growing in arid regions, the flow of CO_2 towards Rubisco is restricted to such a degree that there is a need for a CO_2 concentrating mechanism (CCM), and this is in C_4 and crassulacean acid metabolism (CAM) plants realised as C_4 metabolism. We shall see in the next chapter that marine plants also need a CCM, but this is generally realised in mechanisms of HCO$_3^-$ utilisation rather than C$_4$ metabolism.

We consider the above principal treatment of photosynthesis *per se* as adequate for Chapter 7, i.e. the one that treats marine plants specifically. There are numerous books on photosynthesis that detail the process further. Among them, it should now be clear that we have a favourite, which we therefore highly recommend: the book by Robert Blankenship entitled Molecular Mechanisms of Photosynthesis.

The seagrass *Posidonia oceanica* in the Mediterranean. Photo by Mats Björk.

Chapter 7

Acquisition of carbon in marine plants

The light reactions of marine plants are similar to those of terrestrial plants (see Chapter 5), except that pigments other than chlorophylls *a* and *b* and carotenoids may be involved in the capturing of light (see Chapter 4) and that special arrangements between the two photosystems may be different (e.g. in plants growing under a different spectrum of light at depth, see the forthcoming Section 9.5). Similarly, the CO_2-fixation and -reduction reactions are also basically the same in terrestrial and marine plants. Perhaps one should put this the other way around: Terrestrial-plant photosynthesis is similar to marine-plant photosynthesis, which is not surprising since plants have evolved in the oceans for 3.4 billion years and their descendants on land for only 350–400 million years. During the latter time, one main evolutionary feature of terrestrial plants was the formation of a waxy cuticle so as to restrict water loss. At the same time, stomatal openings evolved so as to ascertain a flux of CO_2 from the atmosphere towards the chloroplasts while minimising water loss.

In situations where the CO_2 flux would limit photosynthesis severely, i.e. where the stomata would have to be partly or fully closed during the day so as to limit excessive water losses, C_4 and CAM metabolism evolved as CO_2-concentrating mechanisms (CCM) so as to supply CO_2 to the Calvin cycle. In underwater marine environments, the accessibility to CO_2 is low mainly because of the low diffusivity of solutes in liquid media, and for CO_2 this is exacerbated by today's low (albeit now slowly increasing, see, e.g. Sections 3.2 and 12.2) ambient CO_2 concentrations. Therefore, there is a need for a CCM also in marine plants, as will be described in detail in this chapter. Marine plants have, unlike their terrestrial counterparts, access to external inorganic carbon (Ci) forms other than the CO_2 dissolved in seawater that are present in higher concentrations than the latter, and it will be emphasised from now on that one of those, bicarbonate (HCO_3^-), can be acquired and utilised for their photosynthetic needs in the marine environment. Alternatively, in some cases CO_2 can

Photosynthesis in the Marine Environment, First Edition. Sven Beer, Mats Björk and John Beardall.
© 2014 John Wiley & Sons, Ltd. Published 2014 by John Wiley & Sons, Ltd.
Companion Website: www.wiley.com/go/beer/photosynthesis

per se be actively accumulated towards Rubisco and the Calvin cycle, but in ways that differ from C_4 metabolism or CAM. In any case, it is the importance of active or facilitated transport of Ci from seawater towards the intracellular site of photosynthesis that made us choose to call this chapter of our book 'Acquisition of Carbon in Marine Plants'; indeed, we view the main difference in photosynthesis between marine and terrestrial plants as the latter's ability to acquire Ci (in most cases HCO_3^-) from the external medium and concentrate it intracellularly in order to optimise their photosynthetic rates or, in some cases, to be able to photosynthesise at all.

We know from Section 3.2 that CO_2 dissolved in seawater is, under air-equilibrated conditions and given today's seawater pH, in equilibrium with a >100 times higher concentration of HCO_3^-, and it is therefore not surprising that most marine plants utilise the latter Ci form for their photosynthetic needs. However, we also know from Section 6.1 that Rubisco, the universal enzyme for photosynthetic CO_2 fixation in the equally universal Calvin cycle for CO_2 reduction to sugars uses CO_2 for its carboxylation reaction. Therefore, any plant that utilises bulk HCO_3^- from seawater must convert it to CO_2 somewhere along its path to Rubisco. This can be done in different ways by different plants and under different conditions, but the basic principle of HCO_3^- utilisation in marine plants is outlined in Figure 7.1 (while specific cases will be treated in the forthcoming sections).

7.1 Cyanobacteria and microalgae

Also Rubiscos from the eukaryotic microalgae and, especially, cyanobacteria, show competi-

tion at their active sites between CO_2 and O_2, potentially leading to the inefficiencies in CO_2 fixation, as exacerbated also by photorespiration, described previously (Section 6.1). These Rubiscos, as well as those from many algae, have a range of S_{rel}, $K_m(CO_2)$ and $k_{cat}(CO_2)$ values (see Equation 6.1 in Section 6.1 for these terminologies): some red algal Rubiscos having the highest S_{rel} but relatively low $k_{cat}(CO_2)$, and dinoflagellate L_2 (or Form II) and cyanobacterial L_8S_8 enzymes having the lowest S_{rel} and, where tested, highest $K_m(CO_2)$ and $k_{cat}(CO_2)$ (Table 7.1). Regardless of the absolute values for $k_{cat}(CO_2)$, Rubiscos as a family thus have low specific reaction rates for the carboxylase at CO_2 saturation.[1]

In the late 1970s and early 1980s, work by Aaron Kaplan and Murray Badger, then postdocs in Joe Berry's lab at the Carnegie Institute of Washington, showed that cells of the green microalga *Chlamydomonas reinhardtii* and the cyanobacterium *Anabaena variabilis* showed kinetics of Ci-dependent oxygen evolution indicative of a very high affinity of photosynthesis for CO_2 and that these cells had high intracellular pools of Ci. These observations led to the recognition that because of the extreme constraints regarding Rubisco's carboxylation properties, most cyanobacteria and microalgae had evolved mechanisms to improve the supply of CO_2 to the active site of Rubisco (although an additional role of CCMs as a means to dissipate excess light energy has

---

[1] Partly because of its low affinity to CO_2 (exacerbated by its oxygenase activity), Rubisco has been viewed as the most abundant soluble protein on Earth (and is, thus, a major protein source for vegetarians). Even with the realisation that CCMs somewhat offset the need for its huge quantity, and the recent finding that it is not such a major protein in marine phytoplankton (see the article by Losh, Young and Morel, 2013), John Raven in a commentary in that same issue summarises that Rubisco is, probably, still the most abundant protein on land as well as in the oceans.

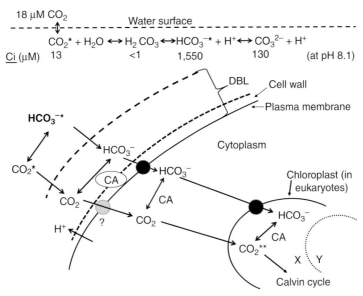

18 µM CO_2

Water surface

$$CO_2{}^* + H_2O \leftrightarrow H_2CO_3 \leftrightarrow HCO_3{}^{-*} + H^+ \leftrightarrow CO_3{}^{2-} + H^+$$

Ci (µM)   13   <1   1,550   130   (at pH 8.1)

Figure 7.1   Basic principles of inorganic-carbon (Ci) **utilisation in marine plants.** An atmospheric CO_2 concentration of 394 ppm (17 µM) is in equilibrium with ~13 µM dissolved CO_2 and ~1,500 µM $HCO_3{}^-$ (depending on temperature and salinity, see Section 3.2) and those two forms of inorganic carbon (Ci, marked with *) are available for marine plants' photosynthetic needs; since the concentration of $HCO_3{}^-$ is ~120 times higher than that of CO_2, most of them prefer the former as their external Ci source. Carbon dioxide is the only form of Ci to be fixed by ribulose-1,5-bisphosphate carboxylase (Rubisco) in the Calvin cycle (marked with **), and therefore $HCO_3{}^-$ has to be converted to CO_2 prior to its fixation and reduction. On the way to CO_2 fixation, Ci has to pass at least three barriers in eukaryotic algae and seagrasses: 1) The diffusion boundary layer (DBL, including the cell wall, leading to the surface of the plasma membrane), 2) the plasma membrane (leading to the cytoplasm) and 3) the chloroplast membranes (leading to the stroma of the chloroplast, X), and it does so either by diffusion of CO_2 or by uptake via transporters (black circles; membrane transport of $HCO_3{}^-$ is only via transporters, and active transport of CO_2 is questionable, thus the question mark and the grey circle). A fourth barrier could be that leading into the pyrenoid, Y (if present inside the chloroplast). Interconversions between CO_2 and $HCO_3{}^-$ within all of these compartments can be catalysed by the enzyme carbonic anhydrase (CA). In the prokaryotic cyanobacteria, compartments X and Y are replaced by the carboxysome. Drawing by Sven Beer.

also been proposed, see Section 11.1). These processes leading to enhanced CO_2 supply are termed CO_2 concentrating mechanisms (CCM). Indeed, species with low S_{rel} Rubiscos would find it difficult to carry out net carbon assimilation under present day CO_2 and O_2 levels without recourse to a CCM. There are several different ways in which a higher CO_2 concentration at the active site of Rubisco can be achieved, and these are dealt with below.

7.1.1 Cyanobacteria

The α-cyanobacteria (those containing a type of Rubisco called Form 1A) of the marine environment are slow growing (perhaps as slow as one division per day) but, because of their high abundance over immense areas, make large contributions to open ocean productivity, especially in low-latitude oligotrophic systems where genera such as *Synechococcus* and

Table 7.1   Selectivity factor (S_{rel})**, half-saturation constant for CO$_2$** ($K_m(CO_2)$) **and CO$_2$-saturated specific rate of carboxylase activity** ($k_{cat}(CO_2)$) **of Rubisco from algae** and higher plants, and the accumulation factor (the ratio between internal and external CO$_2$ concentrations). Based on Raven and Beardall, 2003 and Johnston, 1991.

Rubisco source	S_{rel}	$K_m(CO_2)$ (μM)	$k_{cat}(CO_2)$ (mol CO$_2$ fixed mol active site^{-1} s^{-1})	CO$_2$ Accumulation factor
Dinoflagellates (L$_2$)	37	–		2–26
Cyanobacteria (L$_8$S$_8$)	35–56	105–185	11.4–12	800–900
Green microalgae (L$_8$S$_8$)				
2 species with CO$_2$ concentrating mechanisms	61–83	29–38		5–180
1 species (*Coccomyxa* sp) with diffusive CO$_2$ entry	83	12		~1
Red microalgae (L$_8$S$_8$)				
Porphyridium cruentum	128	22	1.6	3
Cyanidium – 2 species *Cyanodioschyzon*	224–238	6.6–6.7	1.3–1.6	20–80
Diatoms (L$_8$S$_8$)				
3 species with CO$_2$ concentrating mechanisms	106–114	31–36	0.78–5.7	3.5–20
Marine Macroalgae (L$_8$S$_8$)				
Green algae		30–68		
Brown algae		12–60		
Red algae		30–85		
Terrestrial angiosperms (L$_8$S$_8$)				
1 species *(Zea mays)* with a CO$_2$ concentrating mechanism	78	32	4.2	
2 species with diffusive CO$_2$ entry	82–90	10–11	2.9–3.0	

Prochlorococcus can contribute 30–80% of the primary productivity.

The flux of CO$_2$ across the plasma membrane of cyanobacteria is, if present, by diffusion, but on either the cytosolic face of that membrane or on the thylakoid membrane there is an energized conversion of CO$_2$ to HCO$_3^-$ via a NAD(P)H dehydrogenase (Figure 7.2). This then effectively acts like a Ci pump, even though direct active transport of CO$_2$ does not occur. It is possible that some Ci uptake may also be mediated by proteins termed aquaporins. There are two NAD(P)H dehydrogenase systems involved: an inducible, high-CO$_2$ affinity, system at the thylakoid membrane and a constitutive, low-affinity, one (probably at the plasma membrane). Details are illustrated in Figure 7.2 and described fully in the recent reviews by Giordano, Beardall and Raven, 2005, and Price, Badger, Woodger and

Long, 2008. In addition, cyanobacteria can actively take up HCO$_3^-$ from the medium (transport has to be active via transporters as HCO$_3^-$ does not readily cross biological membranes by diffusion). A number of different HCO$_3^-$ 'pumps' have been identified: the genes for one low-CO$_2$ inducible, high-affinity HCO$_3^-$ transporter, BCT1, are found only in freshwater β-cyanobacteria (which contain what is termed Form 1B Rubisco) and are absent from all genomes of marine α- and β-cyanobacteria sequenced so far. A high-affinity, Na$^+$-dependent, HCO$_3^-$ transporter is present in *Synechocystis* 6803 and various other constitutive and inducible transporters have been identified as indicated in Figure 7.2. Again, we refer to Giordano's et al. and Price's et al. reviews for a more detailed discussion.

Bicarbonate (HCO$_3^-$) accumulated in the cytosol via the activity of the various Ci uptake

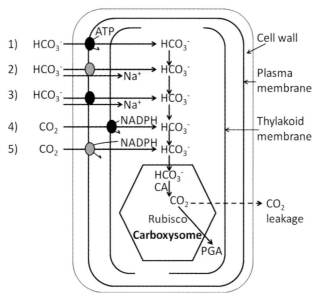

Figure 7.2   A schematic model for inorganic carbon transport in cyanobacterial cells. The 5 major different transport systems are: 1) A low-CO_2 inducible, high-affinity HCO_3^- transporter, BCT1 (only so far found in freshwater β-cyanobacteria); 2) A low-affinity, Na^+-dependent HCO_3^- transporter (BicA) which is nonetheless capable of high flux rates. BicA may be a Na^+/HCO_3^- symporter; 3) A high affinity, Na^+-dependent, inducible HCO_3^- transporter (SbtA). Like BicA, this may be a Na^+/HCO_3^- symporter; 4) A thylakoid-based NAD(P)H dehydrogenase, $NDH-I_3$. This CO_2 uptake system is inducible under low-Ci levels and has a high uptake affinity; 5) $NDH-I_4$ is a CO_2 uptake system based on the plasma membrane, but is constitutive and with a lower affinity than $NDH-I_3$. CA = carbonic anhydrase. Drawing by John Beardall.

systems serves as a substrate for a CO_2 fixation trap where, with the help of the enzyme carbonic anhydrase (CA), CO_2 is thus locally produced within the carboxysome and used by the primary photosynthetic carboxylase, Rubisco. The protein shell around the carboxysome may act as a barrier to prevent CO_2 leakage back into the cytosol, but there is currently little empirical evidence for this. Furthermore, a barrier is not essential because the 3-dimensional diffusion within the carboxysome and consumption by Rubisco will tend to lower the CO_2 concentration towards its shell. A model of carboxysomal function that does not rely on a protein shell has been proposed by Reinhold, Kosloff and Kaplan in their 1991 paper. In all, CCMs in cyanobacteria are highly active and accumulation factors (the internal *vs.* external

CO_2 concentrations ratio) can be of the order of 800–900 (see Table 7.1).

7.1.2 Eukaryotic microalgae

Eukaryotic microalgae have also been shown to use both HCO_3^- and CO_2 as external Ci sources for photosynthesis. In some cases direct transport of HCO_3^- takes place, and this can occur concurrently with active CO_2 transport though, again, evidence for the latter is weak. Bicarbonate use involves the activity of an external, periplasmic (outside the plasma membrane but within the cell wall, i.e. within the diffusion boundary layer, DBL) CA. This enzyme is responsible for dehydration of HCO_3^- in the periplasmic space, thereby

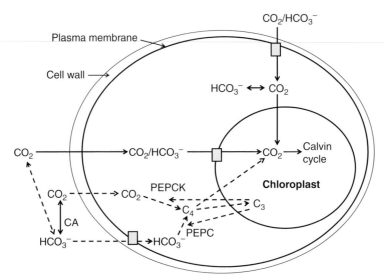

Figure 7.3   A schematic model for inorganic carbon (Ci) **transport and CO$_2$ accumulation processes in eukaryotic microalgae.** The model incorporates the possibilities for Ci transport at the plasma membrane and/or chloroplast envelope, as well as a C$_4$-like mechanism (broken arrows). Such a C$_4$ mechanism could be based on carboxylation mediated by phosphoenolpyruvate carboxylase (PEPC) using HCO$_3^-$ as substrate, or by phosphoenolpyruvate carboxykinase (PEPCK), using CO$_2$ as substrate. CO$_2$ will cross membranes by diffusion, whereas active transport (shown by the shaded boxes) can be of HCO$_3^-$ and, but less likely, of CO$_2$. This model does not incorporate the roles of the various forms of internal carbonic anhydrase (CA) in the different compartments; for this the reader is referred to Murray Badger's paper from 2003. Drawing by John Beardall.

facilitating the supply of CO$_2$ at the plasma membrane and improving the potential for CO$_2$ uptake. (This system is also very common for the marine macrophytes, see Sections 7.3 and 7.4.) Possible mechanisms of Ci transport in eukaryotic microalgae are shown in Figure 7.3.

A range of genes involved in Ci transport have been identified in microalgal groups, including an anion exchange (AE) HCO$_3^-$ transporter, though the latter mechanism is more common in macroalgae (see below). In eukaryotic algae, Ci transport can occur at the plasma membrane, or the chloroplast envelope, or both. In green microalgae such as *Chlamydomonas*, isolated chloroplasts have been shown to possess inducible high-affinity Ci uptake systems as well as a constitutive, low-affinity transporter, resembling the case of

cyanobacteria, thought to be the ancestors of the modern-day chloroplast (see Section 1.1).

In one species of diatom, *Thalassiosira weisflogii*, there is some evidence for the operation of a single-celled C$_4$ pathway, which might operate as a biochemical rather than a biophysical (based on active transport) CCM. This is though still somewhat contentious and the C$_4$ cycle in this diatom may not raise the CO$_2$ levels at the Rubisco active site and may only be involved as a way to dissipate excess light energy.

Coccolithophores can generate CO$_2$ internally as part of the calcification process (see Figure 7.15) and it has been suggested that this could act as a CCM. However, recent data show that although the CO$_2$ from calcification may end up participating in photosynthetic carbon fixation, such a link is not obligatory and calcification does not function as a CCM. Nonetheless, coccolithophores do have CCM activity, but this must be based on the mechanisms for active Ci transport described above.

CCMs in eukaryotic microalgae are not as effective at raising internal CO_2 concentrations as are those in cyanobacteria, but then they do not need to be, given the lower $K_m(CO_2)$ values of their Rubiscos. Nonetheless, microalgal CCMs result in CO_2 accumulation factors as high as 180 (Table 7.1), though for most genera examined values are considerably lower than this.

7.2   Photosymbionts

Marine photosymbionts are those photosynthetic cyanobacteria and microalgae that live within several types of marine invertebrates (see Section 2.3). It has classically been thought that one of the 'advantages' of photosymbionts over free-living cyanobacteria or microalgae is that they would utilise host-respired CO_2. If so, this would alleviate Ci limitations and, perhaps, invalidate the need for a CCM. However, at least for the very common eukaryotic photosymbiont *Symbiodinium* spp. (also called zooxanthellae), rates of photosynthesis *in hospite* (within the host) are generally higher than what could be explained by the supply of respired host-CO_2 only. This is not surprising since the rate of photosynthesis must surpass that of respiration of the holobiont (host plus photosymbiont) if the host's entire nutritional requirements are to be met by the photosymbiont's supply of photosynthates (which is the case for many corals, see Box 7.1). Advances in photosynthetic research of photosymbionts have been gained mainly for the genus *Symbiodinium* in cnidarians (corals and sea anemones) and the giant clam *Tridacna*, and for cyanobacteria in some sponges (see Box 2.2). From those studies it emanates that

Ci from seawater is used as the main source of CO_2 for the photosymbionts' photosynthesis. In the cnidarians, the distance from the host-derived perigleal membrane that surrounds the intracellularly living zooxanthellae to the seawater is short, and some type of diffusional (or possibly facilitated) transport is possible. It is, however, equally possible that the host functions as an energy-driven 'shuttle' of Ci that can increase the internal zooxanthellae Ci concentration to several times that of seawater. In *Tridacna*, such a transport system (towards the, here, intercellular photosymbionts) is more complicated and involves several host compartments until Ci is delivered to the zooxanthellae (see the 2002 paper by Leggat et al. on dinoflagellate symbioses).

Symbiodinium, the main eukaryotic photosymbiont, or at least the one most studied, is a dinoflagellate, and like for other members of this group its Rubisco is of another type than in most plants and cyanobacteria. This type of Rubisco (termed L_2 or 'Form II' Rubisco, see Section 7.1) features a lower level of discrimination between CO_2 and O_2 than the more common type. Also, like for some other microalgae and macroalgae, the $K_m(CO_2)$ of zooxanthellar Rubisco is around 50 µM, while the seawater CO_2 concentration is some 13 µM. These constraints, and the uncertainty that host-respired CO_2 can contribute significantly to their photosynthetic needs, call for a CCM to be present in zooxanthellae, and although their pathway of Ci utilisation has not been clarified yet, this CCM most probably depends on the localisation of Rubisco within the pyrenoid and that of various forms of carbonic anhydrase (CA) in virtually all compartments along the path from the surrounding seawater to Rubisco. If so, and given that CA can only catalyse the otherwise slow inter-conversion between CO_2 and HCO_3^- (but not concentrate one Ci form above what is determined by the pH), it is conceivable that different

Box 7.1   Carbon acquisition in corals

by **Gidon Winters**, The Dead Sea – Arava Science Center, IL (wintersg@adssc.org)

The success of hermatypic (reef-building) corals in oligotrophic waters has been related to their symbiosis with photosynthetic unicellular dinoflagellates, mostly of the genus *Symbiodinium*, commonly known as 'zooxanthellae'. These photosymbionts occur in high densities in the coral host's gastrodermis tissue (~1–5×10^6 algal cells cm^{-2} of coral tissue). Similar to other dinoflagellates, zooxanthellae contain characteristic photosynthetic pigments that include primarily chlorophyll *a*, chlorophyll c_2 and carotenoids such as peridinin. Studied extensively for the last 40 years, it has been shown that hermatypic corals depend on the photosynthesis of their algal symbionts for their main source of organic carbon. Indeed, this 'translocation' process, in which carbohydrates (e.g. glycerol and glucose) and amino acids (e.g. alanine) from the zooxanthellae are translocated to the coral host, is able to meet the entire metabolic requirement of the host and allow coral growth even without the coral practising predation. Additionally, the coupling of zooxanthellae photosynthesis with the host's calcification process has been shown to increase the holobiont's (coral's) calcification rate 2–3 times in light compared to in darkness, a phenomenon known as light-enhanced calcification. Indeed, in addition to the lack of photosynthates to 'feed' the corals, it may be the light-enhanced calcification that limits reef-building (hermatypic) corals to the photic zone.

In the mutualistic symbiosis between corals and their algal symbionts, the host provides its symbionts with inorganic carbon (Ci) and macro-nutrients (mostly nitrogen and phosphorus), the source of which are zooplankton predation and/or the possible digestion of some of the zooxanthellae. It is also assumed that algal symbionts benefit from this symbiosis due to the protection provided by the host since reports on the isolation of free-swimming zooxanthellae from the water column are very scarce.

Relationships between symbiont density and photosynthetic carbon acquisition have been of major interest in coral studies. While Ci acquisition initially increases with higher densities of zooxanthellae, 'self shading' or limited availability of Ci eventually reduces rates of photosynthesis per cell and limits carbon translocation to the host. The fact that, at least over short time periods (days), *in hospite* (within the host) zooxanthellae populations appear to remain stable means either that the division rate is similar to the mortality rate, or that some kind of control of zooxanthellae population density might be in effect, possibly through the digestion of zooxanthellae or by regulating the amount of the nutrients provided to the symbionts through the use of 'host factors'.

The maintenance of an 'optimum' carbon acquisition rate is also evident from studies looking at the how corals photoacclimate to changes in light intensity and spectrum associated with increasing depths (spatial photoacclimation) and the seasonal changes in light intensity experienced in shallow environments (temporal photoacclimation) throughout the year. Increases in chlorophyll density were found in deeper *vs.* shallower corals. Similar differences were also found in shallow growing corals measured in the winter *vs.* summer months. However, while exposure to low-light environments resulted in high chlorophyll densities in both of these processes, their mechanisms were profoundly different: With increasing depths, corals did not differ in zooxanthellae density (which would have probably entailed self-shading effects) but algal cells from deeper corals contained some 4 times more chlorophyll than cells from their shallower counterparts. In contrast, in shallow environments where there would be enough light to compensate for self-shading, corals sampled in the (low-light) winter were found to contain 2–3 times more algae than the same colonies sampled in the (high-light) summer months. These results point to some 'optimum' amount of symbionts that has to be maintained and adjusted according to light intensities and nutrient availability. Too many algal cells could result not only in self-shading (unless light was not a limiting factor), but also in a large carbon cost (due to algal respiration).

Another mechanism to control carbon acquisition in algal symbionts (except for controlling their number and pigment content) would be to control the *in hospite* light environment available to them. Indeed, it has been shown that in shallow corals where light is not a limiting factor, zoox-anthellae are found deeper in the host's tissue compared with cells from deeper corals. Likewise, zooxanthellae from shallow corals are found deeper in the host's tissue compared with the same colonies sampled during the winter months. These studies point to host-mediated mechanisms that control the light environments available to the zooxanthellae. In low-light environments, zooxanthellae are 'allowed' to grow nearer the light surface, thereby making them more efficient in capturing light. In high-light environments, the host tries to minimize the amount of cells, and tries to shade them by pushing them deep into the tissue. Too much light would entail higher rates of photosynthesis, which could also lead to photoinhibition and the development of reactive oxygen species (ROS); ROS would be harmful to both the coral and its symbionts. Another way to control the levels of ROS would be to reduce photosynthetic efficiency and electron-transfer rates during such periods of high light intensities. All these factors would greatly affect rates of photosynthetic carbon acquisition.

pH values in the different compartments play an important role in supplying Ci towards Rubisco. It is also possible that, like other microalgae (see Section 7.1), the CCM is located to the plasma membrane, chloroplast or pyrenoid of the zooxanthellae only. This could especially be the case for the cnidarians where, again, the distance between the seawater and the photosymbionts is small. In any case, a CCM is indicated, e.g. by zoxanthellar pho-tosynthesis being relatively insensitive to O_2. From the few actual measurements available of freshly isolated zooxanthellae it seems that they can maintain up to 10 times the seawater concentration of Ci, showing that indeed they have a functioning CCM.

The form of nutrition supplied to the host by the photosynthetic activity of the photo-symbiont varies between organisms. In corals, glycerol seems to be the main energy-rich compound exported, while mainly sucrose is supplied by the zooxanthellae to the giant clam tissues. In addition, lipids and proteins or, at a minimum, amino acids, are released by the photosymbiotic algae, at least in organisms where the host depends completely on the nutrition supplied by the photosymbiont. More complex interplays between hosts and photo-symbionts include the possibility that the host produces compounds that stimulate the release of photosynthates; those symbiotic interactions will become detailed as other, perhaps young upcoming, scientists tackle the novel and, at least for us, interesting field of photosymbiosis.

7.3 Macroalgae

As in most other marine plants, most macroalgae use mainly HCO_3^- from the external seawater medium for their photosynthetic needs. Again, the 'logical' rationale for this is that HCO_3^- occurs at a >100 times higher concentration than CO_2 in air-equilibrated seawater of pH 8.1. We shall now first examine the evidence for HCO_3^- utilisation in marine macroalgae, and then look at the mechanisms involved in utilising HCO_3^-.

7.3.1 Use of HCO_3^-

The conclusion that macroalgae use HCO_3^- stems largely from results of experiments in which concentrations of CO_2 and HCO_3^- were altered (chiefly by altering the pH of the seawater) while measuring photosynthetic rates, or where the plants themselves withdrew these Ci forms as they photosynthesised in a closed system as manifested by a pH increase (so-called pH-drift experiments). In the latter, in principle, the higher the pH could be driven by photosynthesis, the more efficient was HCO_3^- use since CO_2 concentrations approach zero at pH >9 (see Figures 3.4 and 3.6) and the plants then photosynthesise as a function of HCO_3^- only; CO_3^{2-} is not used since photosynthesis usually ceases at high pH values (pH >10) despite the fact that this Ci form then is present at high concentrations. The reason that the pH in the surrounding seawater increases as plants photosynthesise is first that CO_2 is in equilibrium with carbonic acid (H_2CO_3), and so the acidity decreases (i.e. pH rises) as CO_2 is used up. At higher pH values (above ~9), when all the CO_2 is used up, then a decrease in HCO_3^- concentrations will

also result in increased pH since the alkalinity is maintained by the formation of OH^-; as will be seen in the next section, some algae can also give off OH^- to the seawater medium in exchange for HCO_3^- uptake, bringing the pH up even further (to >10).

An example of experiments in which CO_2 vs. HCO_3^- utilisation in the common green macroalga *Ulva* sp. was elucidated is given in the following: Photosynthesis (i.e. photosynthetic rates) of circularly cut *Ulva* discs was measured in a closed O_2-electrode system, in which pH was also measured simultaneously (see Figure 7.4), as a function of CO_2 and HCO_3^- concentrations that were altered in several ways: In the first approach, an acid (HCl) or a base (NaOH) was injected into the system while monitoring photosynthetic O_2 evolution and pH. The different Ci forms were then calculated at each pH (similarly to what was calculated for a closed system in Figure 3.4) and plotted together with photosynthetic rates as shown in Figure 7.5a. Since 1) the CO_2 concentrations at pH >9 approached zero, 2) photosynthesis at pH 9–10 followed the HCO_3^- concentration and 3) photosynthesis ceased at pH values where the HCO_3^- concentration approached zero (although there was plenty of CO_3^{2-} present, not shown in the figure), photosynthesis of *Ulva* as a function of HCO_3^- concentration could be plotted as in Figure 7.6a (black dots). In a second set of experiments, HCl was added to open beakers containing seawater while monitoring pH, and the beakers were bubbled with air so as to cause the excess CO_2 formed by the other Ci forms (according to equilibrium formula 3.4 in Section 3.2) to equilibrate with the atmosphere in accordance with the curves in Figure 3.5 for open seawater systems. Thus, as seen from the dotted line in Figure 7.5b, the final, air-equilibrated, CO_2 concentration of the seawater was the same at all pH values (11 µM in 1983) while that of HCO_3^- decreased with decreasing pH (and CO_3^{2-} was zero at those

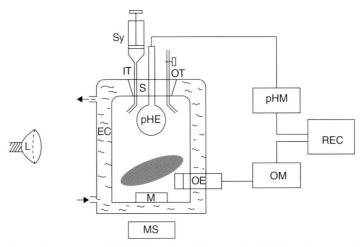

Figure 7.4   Experimental setup used for measuring photosynthetic O$_2$ exchange rates of marine plants (in this case circular thallus discs of *Ulva* such as indicated by the grey ellipse) while changing the inorganic-carbon (Ci) concentration and CO$_2$/HCO$_3^-$ ratio of the synthetic seawater by injecting NaHCO$_3$ or CO$_2$ solutions and acids or bases, respectively, and also measuring the resulting pH. L, light source; EC, external (water-jacketed) chamber for keeping constant temperature by temperature-controlled water circulating in the jacket; Sy, syringe for injecting acids (mostly HCl) or bases (mostly NaOH) through an inlet tube (IT) while releasing overpressure through an outlet tube (OT); OE and OM, O$_2$ electrode and O$_2$ meter, respectively; pHE and pHM, pH electrode and pH meter, respectively; REC, recording device for O$_2$ concentrations and pH (later replaced by computerised data collection); M and MS, magnet and magnetic stirrer, respectively, so as to keep the water and alga moving in the system. Reprinted from: Beer S, Eshel A, Waisel Y, 1977. Carbon metabolism in seagrasses. I. The utilization of exogenous inorganic carbon species in photosynthesis. Journal of Experimental Botany 106: 1180–1189. Copyright (2013), with permission from Oxford University Press.

low pH values). These seawater solutions of different pH values were then used for measuring photosynthesis in the same O$_2$-electrode system as described above, and the results are presented in Figure 7.6a (x dots). Together, these two experiments show that photosynthesis as a function of HCO$_3^-$ concentration saturates at 1000 µM, which is less than the ~1500 µM present in seawater. In order to determine the photosynthetic response to CO$_2$, aliquots of a saturated CO$_2$ solution were injected into a Ci-free synthetic seawater solution in the measuring system, and the results are shown in Figure 7.6b. From this figure, the contribution of the 11 µM CO$_2$ present in the second approach could be determined. More importantly, the figure shows that CO$_2$ saturation was reached at *ca.* 80 µM, which is much higher than the 11 µM present in air-equilibrated seawater at

the time. It can be concluded from these experiments that *Ulva*'s affinity for CO$_2$ is higher than for HCO$_3^-$, but that this fact cannot much 'aid' *Ulva* in normal seawater since the CO$_2$ concentration there is much lower than what this alga would need to saturate its photosynthetic potential. On the other hand, the HCO$_3^-$ concentration of natural seawater is saturating for *Ulva*'s photosynthetic needs despite the fact that the affinity for this Ci form is much lower than for CO$_2$. Thus, *Ulva* is a 'HCO$_3^-$ user' in today's air-equilibrated oceans, and only some 10–20% of its photosynthetic rate could be based on, or explained by, CO$_2$ utilisation. Similarly, most other marine macroalgae are also HCO$_3^-$ users, and especially those living in well-lit waters where irradiance is not a strongly limiting factor for photosynthesis and growth.

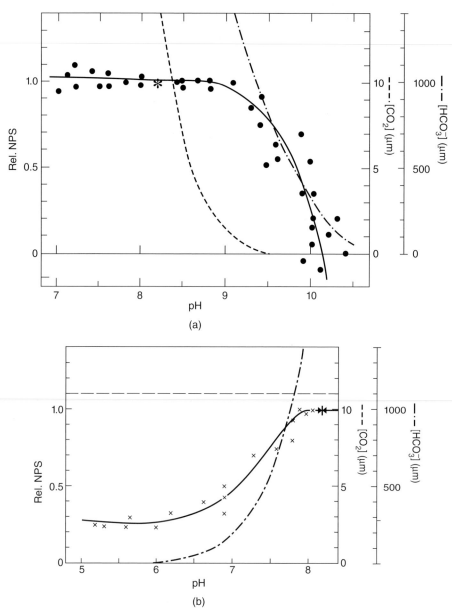

Figure 7.5 (a) **Results of photosynthetic O₂-exchange measurements** (relative 'net' photosynthetic rates (Rel. NPS), full line as supported by data-dots) **of *Ulva* thalli as a function of pH** in the **closed system** depicted in Figure 7.4. Also seen are CO_2 and HCO_3^- concentrations at the different pH values. (b) **Results of photosynthetic O₂-exchange measurements** (relative 'net' photosynthetic rates (Rel. NPS), full line as supported by data dots) **as a function of pH in open beakers** where the seawater had previously been equilibrated with air after adjusting its pH with acid additions. Also seen are CO_2 and HCO_3^- concentrations at the different pH values. Reprinted from: Beer S, Eshel A, 1983. Photosynthesis of *Ulva* sp. II. Utilization of CO_2 and HCO_3^- when submerged. Journal of Experimental Marine Biology & Ecology 70: 99–106. Copyright (2013), with permission from Elsevier.

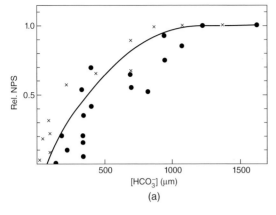

(a)

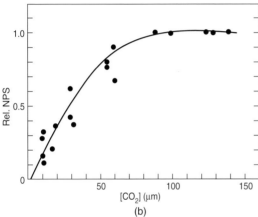

(b)

Figure 7.6 (a) **Rates of relative net photosynthesis** (rel. NPS) **as a function of HCO_3^- concentration of *Ulva* thalli** as derived from Figure 7.5a (black dots) and Figure 7.5b (x-dots). See the text for further interpretations of these results. (b) **Rates of relative net photosynthesis** (rel. NPS) **as a function of CO_2 concentration** as derived from the experiment where a CO_2-saturated solution was injected into the experimental system containing inorganic carbon (Ci)-free synthetic seawater so as to generate the CO_2 concentrations indicated. Reprinted from: Beer S, Eshel A, 1983. Photosynthesis of *Ulva* sp. II. Utilization of CO_2 and HCO_3^- when submerged. Journal of Experimental Marine Biology & Ecology 70: 99–106. Copyright (2013), with permission from Elsevier. (In this book we will call net photosynthesis 'apparent' photosynthesis.)

7.3.2 Mechanisms of HCO_3^- use

The way, or **mechanism,** by which marine macroalgae such as *Ulva* utilise HCO_3^- from the external medium has been elucidated mainly by using inhibitors of carbonic anhydrase (CA)[2]. This enzyme catalyses the otherwise slow equilibrium of the reaction

$$CO_2 + H_2O \leftrightarrow H_2CO_3 \qquad (7.1)$$

within the overall dissolved-Ci equilibria

$$CO_2 + H_2O \leftrightarrow H_2CO_3 \leftrightarrow HCO_3^- + H^+$$
$$\leftrightarrow CO_3^{2-} + 2H^+ \qquad (7.2)$$

When an inhibitor[3] of extracellularly acting (periplasmic) CA was added to the seawater in

[2]Carbonic anhydrase (CA) is a ubiquitous enzyme, found in all organisms investigated so far (from bacteria, through plants, to mammals such as ourselves). This may be seen as remarkable, since its only function is to catalyse the inter-conversion between CO_2 and HCO_3^- in the reaction $CO_2 + H_2O \leftrightarrow H_2CO_3$; we can exchange the latter Ci form to HCO_3^- since this is spontaneously formed by H_2CO_3 and is present at a much higher equilibrium concentration than the latter. Without CA, the equilibrium between CO_2 and HCO_3^- is a slow process (i.e. the hydration of CO_2 and dehydration of H_2CO_3 is a slow reaction with a half-time of up to 0.5 min), but in the presence of CA the reaction becomes virtually instantaneous. Since CO_2 and HCO_3^- generate different pH values of a solution, one of the roles of CA is to regulate intracellular pH (see the last paragraph of the below Personal Note entitled Anion exchange in *Ulva*); another, and here important, function is to convert HCO_3^- to CO_2 somewhere en route towards the latter's final fixation by Rubisco.

[3]The classical CA-inhibitors are ethoxyzolamide and acetazolamide; while the former easily penetrates cell (or plasma) membranes, and thus acts also within most or all compartments depicted in Figure 7.1, the chemical structure of acetazolamide makes it less amenable to such penetration and, therefore, has been considered to affect mainly extracellularly acting CA (at least during short-term experiments). Another membrane-impermeable CA inhibitor can be made of a resin bound to the inhibitor (e.g. dextran-bound ethoxyzolamide) such that the molecule is too large to cross membranes.

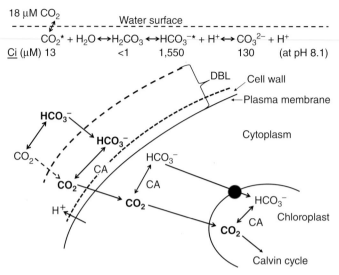

Figure 7.7   Utilisation of inorganic carbon (Ci) in *Ulva* via extracellularly active carbonic anhydrase. See Figure 7.1 for general explanations, abbreviations, etc. Note especially that Ci utilisation here is based on the extracellular (within the diffusion boundary layer, DBL) carbonic anhydrase- (CA)-catalysed conversion of HCO_3^- to CO_2. While the entry of Ci into the cells is by diffusion, its entry into the chloroplasts could be either via CO_2 diffusion or HCO_3^- transport (black circle). See text for details. Drawing by Sven Beer.

which *Ulva* photosynthesised, then photosynthesis ceased almost completely, showing that the fast conversion of HCO_3^- to CO_2 close to the membrane is an important component of HCO_3^--utilisation in this plant.

Based on experiments where photosynthetic rates were measured in the absence or presence of various CA inhibitors, the following is a plausible way in which marine macroalgae such as *Ulva* utilise HCO_3^- (see Figure 7.7): Bicarbonate diffuses from the bulk seawater medium through the diffusion boundary layer towards the cell (or plasma) membrane. Here, while CO_2 diffuses inwards through the cell membrane, as driven by a concentration deficit caused by its intracellular photosynthetic fixation and reduction, extracellularly (or periplasmic-)acting CA catalyses the conversion of HCO_3^- to CO_2 such that CO_2 can be readily resupplied for inward diffusion. It must be remembered that while CA causes the instantaneous conversion of HCO_3^- to CO_2, it

can only replenish that CO_2 that is out of equilibrium with HCO_3^-, e.g. as the former diffuses into the photosynthesising cells, i.e. it cannot increase the CO_2 concentration above what is dictated by pH. From this it follows that in order for the CO_2 concentration to be above the *ca.* 13 µM in equilibrium with the other bulk Ci forms, this would require a lower pH in the diffusion boundary layer. Such lower pH values close to the cell membranes of marine plants are possible based on membrane-located H^+ pumps, and it is assumed that they must be active in order to generate high CO_2 concentrations for inward diffusion, which would explain the high photosynthetic rates of this plant.

Many other marine macroalgae feature the same kind of HCO_3^- utilisation as described above for *Ulva*, i.e. the conversion of HCO_3^- to CO_2 within the diffusion boundary layer outside the plasma membrane. These include numerous green algae, and, e.g. the red alga *Gracilaria* and the brown alga *Fucus* (see

Box 7.2   The *Ulva* dilemma: pH increase or decrease?

by **Alvaro Israel**, Israel Oceanographic and Limnological Research, Tel Shikmona, IL (alvaro@ocean.org.il)

During the race of elucidating the modes of inorganic carbon utilisation in seaweeds, back early in the 1990s when I was a graduate student in Sven's laboratory, we developed a novel way for measuring pH on *Ulva* thallus surfaces, within diffusion boundary layers (DBL), using a flat-tipped pH electrode (intended for pH measurements on the surface of cheeses). Lights on, electrodes in place, and to our surprise the response occurred almost at once: within a few minutes the pH within the DBL of *Ulva* thalli **increased** from an average of 8.2 to close to, or above, 10 (see Figure 7.8b). With a second pH electrode placed a few cm aside we could follow the subsequent rise in pH also in what we stylishly called the 'surrounding media' of *Ulva*. These results were part of the, at least temporary, 'proof' that *Ulva* excreted OH^- (in a later discovered mechanism of exchange with HCO_3^-, see above).

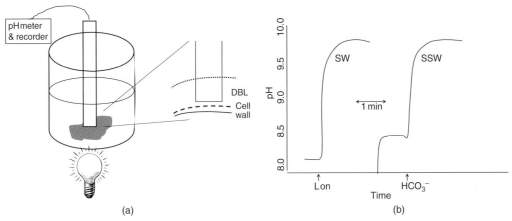

(a)   (b)

Figure 7.8   Surface pH measurements on HCO_3^--transporting *Ulva*. (a) Application of a flat-tipped pH electrode to an *Ulva* thallus (grey blob) such that it measures within the diffusion boundary layer (by touching the cell wall). (b) The resulting pH response in normal seawater (SW) when turning on the light and in synthetic seawater (SSW) when adding $NaHCO_3$. See text for further explanations. Adapted from: Beer S, Israel A, 1990. Photosynthesis of *Ulva fasciata*. IV. pH, carbonic anhydrase and inorganic carbon conversions in the unstirred layer. Plant Cell and Environ. 13: 555–560. Copyright (2006), with permission from John Wiley and Sons.

Later, as an independent researcher, Aaron Kaplan and I encouraged a young graduate student, Tami Beer (no relation to Sven), to repeat the experiments with the flat-tipped pH electrode and complement them with measurements of CO_2 fluxes in and out of *Ulva* thalli using a mass spectrometer (which Aaron possessed). This may have been spurred by Aaron's et al. previous findings that some marine cyanobacteria could expel CO_2 while photosynthesising (see Box 11.1). To Tami's and Aaron's delight, but to my dismay, yes, the pH did increase, but this was followed by a pH **decrease** after a few minutes, contrary to our earlier findings (see the previous paragraph). The decrease in pH could be explained by H^+ extrusion into the DBL, as corroborated by the formation of CO_2 within the DBL, or the CO_2 could have been formed intracellularly by HCO_3^- taken up by the photosynthesizing cells (which would also explain a pH decrease). In any case, *Ulva* did not behave here as in previous experiments. Perhaps *Ulva* can have a third way of utilizing Ci, illustrating that this seaweed has a large plasticity regarding its options for Ci utilisation? Or perhaps this species employs a different way of utilizing Ci than other species? If so, others need to find out whether *Ulva fasciata* increases the pH of the DBL while *Ulva rigida* lowers it!

Box 7.3). There are also several variations on this theme, including higher or lower degrees of involvements of H^+ pumps, especially in the brown algae (again see Box 7.3). Several red algae, and especially those that grow under low irradiances (often the deep-growing ones) lack a system for HCO_3^- use and rely on inwards diffusion of bulk CO_2. The rationale for this is that light, rather than Ci, is the limiting factor for photosynthesis and growth, but more about that in Box 7.3 and Section 9.4.

In the 1990s, Sven Beer and co-workers aimed at showing just how low the pH could be in the diffusion boundary layer of *Ulva* so as to provide enough CO_2 for its effective inward diffusion towards the chloroplasts; *Ulva* features very high photosynthetic rates, and so it was envisioned that the pH would be low so as to generate lots of CO_2. This was done by placing a flat-tipped pH electrode on the surface of an *Ulva* thallus placed on the bottom of a glass flask and illuminating it from below (Figure 7.8a). They found, to their dismay, that not only did the pH within the diffusion boundary layer of *Ulva* not decrease, but it increased to values of ~10 within seconds of illuminating the thalli (see Figure 7.8b). This finding was not expected since the type of HCO_3^- utilisation described above would need at least parts of the algal surface to have a pH of <8 (again, so as to generate a higher CO_2 concentrations than that of air-equilibrated sea-

water). On the contrary, at such high pH values as generated by this *Ulva*, the CO_2 concentration in equilibrium with ionic Ci forms would be so low (nanomolar, nM, levels) as not to make inwards diffusion possible. This finding, however, was the first sign of another type of HCO_3^- utilisation in *Ulva* based on direct HCO_3^- uptake (see Figure 7.9). The model for this kind of HCO_3^- utilisation was found to depend on an alleged membrane transporter of HCO_3^- called an anion exchange protein (AE, see SB's Personal Note below): As HCO_3^- moves inwards with a concentration gradient, the AE shuttles another anion outwards. This other anion appeared to be hydroxyl (OH^-), which is a product of converting HCO_3^- to CO_2 in the cytoplasm (Figure 7.9). An interesting analogue to the AE found in *Ulva* is also found in red blood cells, where the transporter exchanges HCO_3^- with Cl^- (see, again, the below Personal Note).

The model of HCO_3^- transport into *Ulva* cells in exchange for OH^- has been challenged by one of its co-finders, Alvaro Israel, who later found that pH increased only transiently when illuminating *Ulva* thalli. In order to shed more light on this, he has kindly agreed to write a box on the subject (see above). As in many areas of biology, there may be two (or more) sides of a suggested mechanism, which may depend on the way they were elucidated or, more hopefully, on different environmental

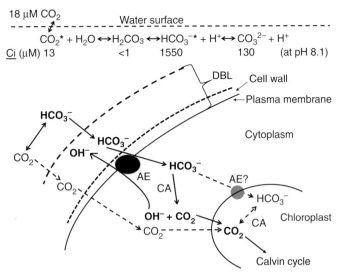

$$18 \ \mu M \ CO_2$$

Water surface

$$CO_2^* + H_2O \leftrightarrow H_2CO_3 \leftrightarrow HCO_3^{-*} + H^+ \leftrightarrow CO_3^{2-} + H^+$$

$Ci \ (\mu M)$ 13   <1   1550   130   (at pH 8.1)

Figure 7.9   **Utilisation of inorganic carbon** (Ci) **in** *Ulva* **via a transmembrane anion exchange protein.** See Figure 7.1 for general explanations, abbreviations, etc. Note especially that Ci utilisation here is based on the transport of HCO_3^- across the cell membrane as mediated by an anion exchanger (AE) (while the contribution of CO_2 fluxes from the outside is minor, if at all existing, as is HCO_3^- transport into the chloroplast, broken arrows). See text for details. Drawing by Sven Beer.

Personal Note: Anion exchange in *Ulva*

In the early 1990s, the technician Zivia Drechsler and the PhD student Rajach Sharkia performed photosynthetic experiments in my (SB's) laboratory, the objective of which was to explain the mechanism of Ci utilisation in a form of *Ulva* that showed high rates of photosynthesis at high pH values (i.e. HCO_3^- to CO_2 conversions in the diffusion boundary layer was unlikely to be facilitated by the efflux of protons, but, rather, some form of HCO_3^- uptake was envisioned, see the text and Alvaro Israel's box above). In doing so, they found that photosynthesis at pH ~10 was sensitive to the anion-exchange protein inhibitor 4,4'-diisothiocyanostilbene-2.2'disuphonate (DIDS). In the literature, we found that DIDS had been used for inhibiting HCO_3^- transport via an anion transport/exchange protein, or anion exchanger (AE) for short, in red blood cells (RBC); and, incidentally, the scientist that had been instrumental in elucidating the mechanism for anion exchange in human blood cells was just 'around the corner'. So I drove up to the Hebrew University in Jerusalem in order to consult with the biochemist Ioav Cabantchik about this strange coincidence of DIDS-sensitive HCO_3^- exchange in RBCs on the one hand and DIDS-sensitive photosynthesis in *Ulva* on the other. One of Yoav's first questions was "*Ulva*? Alga? Sounds fascinating, but how could this anion exchange protein arrive from red blood cells of mammals to the algae of the oceans?" After pondering this for a while, we came to the conclusion that the question should be put the other way around since algae were present in the oceans at least a million years before the mammals even had the faintest thought of evolving. This meeting was the beginning of a fruitful co-operation in which we showed that DIDS could competitively inhibit HCO_3^- uptake in *Ulva*, that this uptake was based on an AE protein and that this AE extracted from *Ulva* cell membranes showed such likeliness to the human AE (termed AE1) that it was recognised by monoclonal (highly specific) antibodies against the human AE1. In conclusions then, while the human AE1 is important for transporting HCO_3^- out and in of RBCs, in *Ulva* it is important in order to 'open up' its cellular gates to the influx of the prevailing Ci form in the oceans, HCO_3^-.

For those interested, the importance of AE1 in RBCs is as follows: As CO_2 is released by cellular respiration, it would acidify the blood plasma to for us dangerously low levels. Instead, the CO_2 diffuses into the RBCs, where it is converted to HCO_3^- as catalysed by carbonic anhydrase (CA). This HCO_3^- can now be exchanged outwards to the plasma, through the transporter AE1, against an influx of chloride (Cl^-). Since HCO_3^- is more of a buffer than dissolved CO_2, the pH of the plasma can thus be maintained at tolerable levels. Near the lungs, a reciprocal process takes place: As CO_2 is exhaled, HCO_3^- can now enter the RBC via AE1 and, there, be reconverted to CO_2, which can diffuse out through the RBC membrane and, subsequently, out of the lungs.

—Sven Beer

Box 7.3   Inorganic carbon utilisation in brown and red macroalgae

By **Lennart Axelsson**, Kristineberg Marine Station, SE (alglax@telia.com)

When I obtained a research position at Kristineberg Marine Biological Station (on the Swedish west coast) in 1980, one of my first tasks was to construct a system for long-term measurements of photosynthetic rates (O_2 production and CO_2 assimilation) under constant conditions. Using this system to evaluate the capability of macroalgae to adapt their mechanisms for inorganic carbon (Ci) uptake to high-pH conditions (i.e. shortage of CO_2, such as, e.g. can occur in calm or enclosed areas like rockpools), I discovered the capability of a number of **green macroalgal species** (including *Ulva lactuca*) to gradually improve their photosynthetic and Ci utilisation capability over a period of about 12 h. (This capability was later shown to depend on the formation of an anion exchange protein that could facilitate the direct uptake of HCO_3^- into *Ulva*'s photosynthesising cells, see earlier in this chapter.) No such capability was apparent in red or brown macroalgae, suggesting that inducible ways of changing the mode of Ci utilisation as a response to environmental change was only present in the green macroalgae (or in microalgae and cyanobacteria in which a CO_2 concentrating mechanism, CCM, could be induced by growth under low-CO_2 conditions). There was, however, a surprising response in some of the **brown macroalgae** upon transfer to high-pH conditions: These algae initially featured a high rate of O_2 production, which was not matched by any concomitant CO_2 uptake. This high O_2 production lasted for 1–2 h, and then gradually (during 4–6 h) decreased to about 1/3 of its initial rate. During the latter period, the uptake of Ci was about the same, matching the steady-state O_2 production at the end of the measurements. Such a result suggested the presence of an internal pool of loosely bound CO_2, which could be used upon shortage of an external substrate, similarly to what occurs in Crassulacean acid metabolism (CAM) of some higher plants (see Section 6.2). This 'photosynthetic buffer' appears now to be confined to the brown algal family Fucaceae, and was most prominent for species adapted to growth high up in the intertidal where they are restricted to aerial CO_2 for prolonged time periods (e.g. *Fucus spiralis* and *Pelvetia canaliculata*) and where, thus, their restricted use of such external CO_2 could be complemented by the internal Ci pool as follows:

Brown macroalgae differ from red and green macroalgae in having light-independent CO_2 fixation (sometimes referred to as 'dark CO_2 fixation'). In this process, phosphoenolpyruvate (PEP) and CO_2 (stemming either from external Ci, respired CO_2, or both) is transformed to oxaloacetate (OAA) in an ATP-consuming reaction:

$$PEP + ATP + CO_2 \rightarrow OAA + ADP \qquad (7.3)$$

OAA is unstable but is, like in C_4 and CAM plants, converted to, and stored as, either malate or aspartate. Based on the measurable changes in titratable acidity, members of Fucaceae thus have a fairly high pool of malate and/or aspartate which can supply CO_2 to their photosynthetic CO_2-fixation and -reduction in the light, and this can enhance their photosynthetic rates:

$$OAA + ADP \rightarrow PEP + ATP + CO_2 \qquad (7.4)$$

Although these algae have a relatively high rate of CO_2 absorption in air (expressed per surface area), they do have a thick thallus that can extend the period of emersed photosynthesis (before they dry out), but which decreases their ability to absorb CO_2 per fresh weight. Thus, 'spreading out' Ci acquisition into the night confers an advantage to them. In darkness, the light-independent CO_2 fixation never exceeds respiration; it amounts to about 50–80% of respired CO_2 as verified by gas-exchange measurements. This is apparently enough to reload the CO_2 buffer in one diurnal cycle.

While the CO_2-buffer mechanism in Fucaceae appears to be an adaptation to air exposure during emersion, many brown algae have a specific mechanism for utilisation of the HCO_3^- pool in seawater. This mechanism, which is prominent in *Laminaria* species (e.g. *Laminaria saccharina*), depends on the excretion of protons (H^+) as well as extracellularly active, plasma-membrane-bound, carbonic anhydrase (CA). This mechanism is therefore almost completely inhibited by acetazolamide (AZ, an inhibitor of external CA), but also by high concentrations (>50 mM) of proton buffers such as TRIS or Tricine (that dissipate any acid regions created by H^+ excretion). (The mechanism is similar to that of some seagrasses, see Section 7.4.) To maintain the overall alkalinity of the plant, the H^+ excretion must be balanced by H^+ uptake (or OH^- excretion) and the two processes must be spatially separated such that local zones of low pH are created. These acid zones have external CA activity, resulting in a rapid generation of a high concentration of CO_2 from HCO_3^-, the former of which can diffuse through the cell membrane and be used for photosynthesis. For this to work, the high-pH zones must be devoid of carbonic anhydrase; thus the comparatively slow rate of the spontaneous hydration of CO_2 to HCO_3^- is maintained in the high-pH zones. A plausible model of the mechanism, with the acid zones and the CA located between adjacent epidermal cells, is presented in Figure 7.10.

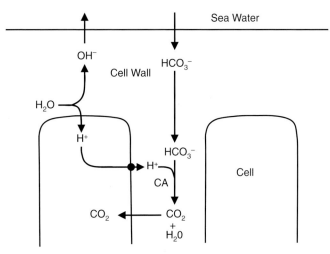

Figure 7.10   A tentative **model for the HCO_3^- utilisation mechanism depending on microzones of low pH**. External carbonic anhydrase (CA) is located outside the plasma membrane between adjacent epidermal cells, where also proton pumps are located (black dot). The uptake of protons (H^+) via the outward located (distal) part of the plasma membrane is assumed to occur via ion exchange. Drawing by Lennart Axelsson.

This mechanism of HCO_3^- utilisation is also operating in members of Fucaceae, but is less efficient at high pH than the direct-HCO_3^- uptake mechanism described for *Ulva* when in the anion-exchange (or AE) mode as brought about by high-pH seawater (see earlier in this chapter). This is probably due to the difficulty of maintaining a low enough pH in the acid zones when the seawater pH is high. For most macroalgae possessing such an H^+-extrusion-enhanced means of HCO_3^--utilization, this mechanism ceases to function at a pH of around 9.6, while this pH is close to optimal for the functioning of the HCO_3^- uptake mechanism of *Ulva*.

If Fucacean algae are trapped in pools of seawater during low tide, their photosynthesis results in a gradual increase in pH of the pool-water. This, in turn, results in a decrease in the efficiency of the 'acid zone' mechanism described above. As pH approaches 9.2, these algae initiate a net excretion of H^+, thus preventing a further increase in pH. They can thus continue their HCO_3^- utilisation at a comparatively high rate until all Ci in the pool is absorbed, or at least until their internal pool of organic acids has been deprotonated. Obviously, then, the CO_2-containing organic- or amino-acid pool of this specific adaptation has a versatile function. Such a mechanism would function well in rockpools within the tidal zone, where the water is replaced regularly and the H^+ pool thus can be reloaded.

Photosynthetic measurements on **red macroalgae**, using AZ as an inhibitor of externally active CA and proton buffers for dissipating acid zones, reveal four different groups using different means of HCO_3^- utilisation. In many subtidal species, e.g. *Delesseria sanguinea* and *Phycodrus rubens*, passive diffusion of CO_2 from the external medium across the plasma membrane is evidently enough to supply photosynthesis. As photosynthesis is inhibited by inhibitors of external CA, this enzyme is capable of assuring a high enough, with seawater equilibrated, concentration of CO_2 just outside the cell membrane. This group of red algae grows at depths where light is more limiting for photosynthesis than the CO_2 availability, and it does not use the sparse energy available to increase the concentration of CO_2 around Rubisco (see also Section 9.4 where the lack of a CCM in low-light growing red algae is discussed further). Typical of this group is a large surface to volume ratio, enhancing both light utilisation and CO_2 uptake.

For red algae with thicker thalli and growing in regimes of higher irradiances, passive CO_2 absorption is not sufficient. Many of those species are strongly inhibited by both AZ and TRIS buffer, thus indicating the operation of acid zones and external CA in a mechanism similar to that described above for the brown alga *Laminaria saccharina*. There are two further experimental findings supporting this mechanism (both experiments were performed in a closed volume): In the first experiment, the algae were placed in a buffered medium, preventing the formation of acid zones. Adding CO_2 resulted in a transient increase in the O_2 production during a short period, until the spontaneous conversion of CO_2 into HCO_3^- was completed (about 1 to 2 min). If CA was added to the medium prior to the supply of CO_2, this transient increase did not occur (because HCO_3^- was formed instantly). If instead AZ was added, the transient became more pronounced. By comparing the two transients (with and without AZ) it was possible to estimate the part of the cell membrane covered with external CA and thus preventing the inward diffusion of CO_2. For *Laminaria saccharina* this value was 40% and for the red alga *Gracilaria tenuistipidata* 35%. As a comparison, *Porphyra* sp. could feature a value of 100% CA coverage. In the second experiment, the pH and the partial pressure of CO_2 in the surrounding seawater was measured concomitantly during a dark/light transition. Any presence of a CO_2 partial pressure above equilibrium indicated CO_2 excretion and could be detected as a decrease in the partial pressure upon the addition of CA to the medium. During respiration in the dark, the partial pressure of CO_2 was slightly above the equilibrium concentration. Upon irradiation, there was a lag phase of *ca.* 5 min, after which photosynthesis started (as indicated by the increase in pH). As photosynthesis started, the alga started to excrete CO_2 at a higher rate (about 2–3 times higher than during respiration). Similar results were obtained for a red (*Gracilaria tenuistipidata*) and a brown (*Laminaria saccharina*) species. Such data are in agreement with a mechanism using acid zones, the latter indicated also by the fact that the addition of a proton buffers stopped photosynthesis by neutralising those zones (in which case also the CO_2 excretion ceased completely). It is worth mentioning that any excretion of CO_2 during photosynthesis provides clear evidence of HCO_3^- use (see an analogous feature for some cyanobacteria and microalgae in Box 11.1). A large number of red algae, perhaps a majority, appear to depend on acid zones for their HCO_3^- utilisation.

Intertidal species of red algae like *Porphyra* are not inhibited by buffer but are heavily inhibited by AZ. Their HCO_3^- utilisation is assumed to be active, but the precise mechanism is not clear. At low Ci concentrations, the addition of TRIS buffer actually results in a pronounced (100%) increase in photosynthesis. It has been suggested that the mechanism for HCO_3^- utilisation in *Porphyra* involves active transport of CO_2 across the cell membrane. Many *Porphyra* species contain pyrenoids in the middle of their chloroplasts. While it has not been shown experimentally, it is possible that those pyrenoids have a CCM function similar to that described for the carboxysomes of cyanobacteria (see Section 7.1). In that case, CA-enhanced passive transport of CO_2 across the plasma membrane would probably be sufficient to supply CO_2 to the photosynthetic CO_2-fixation and -reduction system.

So far one species of red algae, *Palmaria palmata*, is singled out by being unaffected by both AZ and buffer, and it may thus depend on direct HCO_3^- uptake. This alga has been moved to a specific systematic group, Palmariales, and the relevance of this might be supported by its obviously unique mechanism for HCO_3^- utilisation. It would be interesting to examine other members of this systematic group as well – an interesting and rewarding task for upcoming scientists!

---

conditions prevailing during the growth of the organism or during the experimental conditions, or by a mistaken species identity such that two different species were worked on.

We have seen that *Ulva* (within the same species!) can feature two kinds of HCO_3^- utilisation: 1) conversion of HCO_3^- to CO_2 in the diffusion boundary layer as catalysed by CA (therefore here termed the **CA-*Ulva***), followed by CO_2 diffusion into the photosynthesising cells, and 2) direct uptake of HCO_3^- via an anion exchanger (AE, therefore here termed the **AE-*Ulva***). First, then, it should therefore be distinguished that HCO_3^- **utilisation** is not necessarily equal to HCO_3^- **uptake**: in case 1) it is not, while in case 2) it is. Secondly, and more importantly, it can be asked why there are two ways of HCO_3^- utilisation in the same *Ulva* species, and which one works when? The type of HCO_3^- utilisation present in CA-*Ulva* was first described in the early 1980s in the PhD research of Mats Björk. Because he worked in Uppsala at the time, the algae used in his experiments were collected in the cold (<20 °C), low-irradiance (<1000 µmol photons m^{-2} s^{-1}), waters of Sweden; such conditions **are not** conducive to high rates of photosynthesis. On the other hand, the way of direct HCO_3^- uptake present in AE-*Ulva* was elucidated later in Sven Beer's laboratory in Israel, and those algae were collected in the warm (>25 °C), high-irradiance (>1500 µmol photons m^{-2} s^{-1}), waters of the Eastern Mediterranean; such conditions **are** conducive to high rates of photosynthesis. For Mats, the stark discrepancy between his findings and those of Sven were initially disheartening (see his Personal Note below), but with the help of Lennart Axelsson it was elucidated later that *Ulva* could alter its mode of HCO_3^- utilisation depending on environmental conditions: One outcome of high rates of photosynthesis in tropical climates is that the pH in the surrounding water, and certainly in the diffusion boundary layer of macroalgae, can be very high, meaning that the CO_2 concentration there is virtually nil and HCO_3^- (and CO_3^{2-}) is the main Ci form present there. Indeed, Axelsson and co-workers found that CA-*Ulva* could be induced to switch to the AE type by exposing it to high pH for some 24 h. The latter type of HCO_3^- utilisation is more efficient than the CA-mediated one, and may be important in order to achieve high rates of photosynthesis under conditions conducive to such high rates (high temperatures and high irradiances), which would generate high pH values where CO_2/HCO_3^- ratios are low, HCO_3^- to CO_2 conversions would be futile and HCO_3^- uptake would be the best solution to supplying large amounts of CO_2 to the plant. The advantages of this (and other) adaptation to certain environmental conditions will be expanded upon and quantified later, e.g. in Section 11.3.

Personal Note: Mats *vs.* Sven

I (MB) had been working on carbon acquisition in algae for my PhD, and come to the conclusion that HCO_3^- was **converted extracellularly** to CO_2 as catalysed by carbonic anhydrase (CA) in the cell wall, which was then followed by CO_2 influx. Contrary to that concept, and very irritating to me as a young researcher that had to be right, this Sven Beer was at the same time describing HCO_3^- **uptake** in *Ulva* by a direct process involving an anion exchange protein. Even more annoying was that I could not really find any flaws in his analyses of his results, while at the same time I had to trust my own.

I eventually met Sven in 1993 at a workshop on carbon acquisition strategies in marine plants, and we did some exercises together and started discussing, while neither of us admitted to being wrong; we still had a good time and as a result I said that I would find the funds to come to his laboratory in Tel Aviv, where we could try to sort out together the enigmatic carbon strategies of the *Ulva*s. This was 20 years ago, and it was the start of a long friendship. Sven and I still work together, and of course we in the end could show that *Ulva* could feature **both** an extracellular, carbonic anhydrase-mediated, CO_2 uptake, as well as direct uptake of HCO_3^- via a transporter protein... it all depended on the external conditions at which *Ulva* photosynthesised.

—Mats Björk

7.3.3 Rubisco and macroalgal photosynthesis: The need for a CO_2 concentrating mechanism

When there is enough light and nutrients, then the successful photosynthesis (i.e. rates being high enough to support growth) of marine plants depends on efficient fixation of CO_2 by Rubisco in the first step of the Calvin cycle. If we consider that photosynthesis of most terrestrial plants (i.e. the C_3 plants, see Chapter 6) is limited by today's atmospheric CO_2 concentration, then the marine plants, if utilising dissolved CO_2 only, should be even more limited by this Ci source since the diffusivity of solutes such as CO_2 in liquid media is some 4 orders of magnitude lower than in air. Given this constraint, however, the theoretical possibility exists that Rubiscos of marine plants have a much higher affinity for CO_2 (lower $K_m(CO_2)$) than those of terrestrial plants, which was shown above not to be true for microalgae and,

especially, for cyanobacteria. But what about macroalgae? A compilation of $K_m(CO_2)$ values of various macroalgal Rubiscos ranges from 40 to 90 µM, with an average value of 50 µM among 7 species (see Table 7.1). If we confer this value to the 'average Rubisco macroalga' (which, by coincidence, happens to be *Ulva*, having a Rubisco $K_m(CO_2)$ of ~50 µM), it means that macroalgae perform half of their photosynthetic capacity at 50 µM CO_2 (Figure 7.11). Compared to the 13 µM present in today's oceans, this figure shows that virtually no photosynthetic CO_2 fixation could occur in these plants if there was no **CO_2 concentrating mechanism** (CCM) present. Thus, by virtue of necessity, marine macroalgae must possess a CCM in order to sustain high enough photosynthetic rates, at least under conditions where temperature and, especially, light is not a strongly limiting factor.

The presence of a CCM in marina macroalgae was in the early 1980s evidenced by their

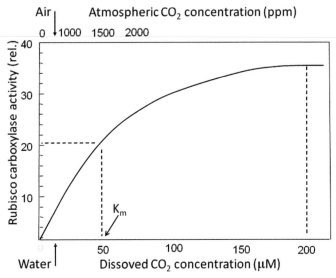

Figure 7.11   The carboxylase activity of 'typical macroalgal' Rubisco, having a $K_m(CO_2)$ of 50 μM (where the reaction rate is half the maximal one, marked on the lower x-axis), as a function of dissolved CO_2 concentrations (lower x-axis, assuming that the intra-chloroplast CO_2 concentration in the stroma of the photosynthesising cell is in equilibrium with that of the atmosphere, upper x-axis). Today's atmospheric (394 ppm) and water-equilibrated (at 20 °C, 13 μM) CO_2 concentrations are indicated by the down- ('Air') and up-ward ('Water') arrows, respectively. The saturating CO_2 concentration (200 μM) is marked by the right vertical broken line. Drawing by Sven Beer.

low CO_2 compensation points and photosynthetic rates not being sensitive to O_2 (i.e. the biochemical pathway of photorespiration was, if not lacking, repressed). Because these features are also typical of terrestrial C_4 plants, it was then assumed that the mechanism behind the lack of photorespiration also in macroalgae was due to an operational C_4 cycle. However, ^{14}C-pulse-chase experiments (similar to the ^{14}C-pulse experiment described in Box 6.1) have shown that, with very few (i.e. hitherto one, a green alga of the genus *Udotea*) exceptions, marine macrophytes are not C_4 plants. Also, while a CAM-like feature of nightly uptake of Ci may complement that of the day in some brown algal kelps, this is an exception for a few specific groups of algae (see Box 7.3) rather than a rule for macroalgae in general. Thus, virtually no marine macroalgae are C_4 or CAM plants, and instead their CCMs

are dependent on HCO_3^- utilisation, which brings about high concentrations of CO_2 in the vicinity of Rubisco. In *Ulva*, this type of CCM causes the intra-cellular CO_2 concentration to be some 200 μM, i.e. ~15 times higher than that in seawater. Under such conditions, *Ulva*'s Rubisco would be saturated by CO_2 (see Figure 7.11) and this alga could, thus, express its high photosynthetic potential leading to high growth rates under favourable growth conditions.

7.4   Seagrasses

Seagrasses also use HCO_3^- from the seawater medium as the principal source of Ci for their photosynthetic needs, and the mechanisms involved are similar to those of most macroalgae. It has been noted that HCO_3^-

utilisation in seagrasses is less efficient than in algae, causing this plant group not to be saturated by today's seawater Ci composition, and this has been seen in connection with the fact that seagrasses developed 'only' some 90 million years ago (as opposed to several hundred million years ago for the macroalgae). However, this notion of a deficient HCO_3^- utilisation system in seagrasses is now changing (see below and in Section 8.4) as new, non-invasive (read: PAM fluorometric, see Section 8.3), methods for measuring photosynthetic rates *in situ* develop.

7.4.1 Use of HCO_3^-

Evidence for HCO_3^- utilisation in seagrasses stems originally from experiments in the late 1970s by SB in which CO_2/HCO_3^- ratios were changed in enclosed systems containing sec-

tions of leaves while measuring photosynthetic rates as O_2 evolution. Those systems were basically the same as depicted for macroalgae in Figure 7.4. Figure 7.12 shows an example of the results of such an experiment: As pH increases, CO_2 concentrations decrease and become virtually nil at pH >8.5. Therefore, photosynthesis measured above that pH is dependent only on HCO_3^-, the concentration of which decreases at high pH values as CO_3^{2-} is formed (also here CO_3^{2-} cannot be used by the plants). At lower pH values, photosynthetic rates increase as CO_2 is released (from the ionic Ci forms) into the closed measuring system. Calculations of the different Ci-forms at the various pH values show that photosynthesis saturates at ~0.5 mM (500 μM) HCO_3^-, i.e. the present-day seawater HCO_3^- concentration of ~2 mM is saturating for photosynthesis, while the corresponding CO_2 concentration (~13 μM) is far from saturating. However, as is evident

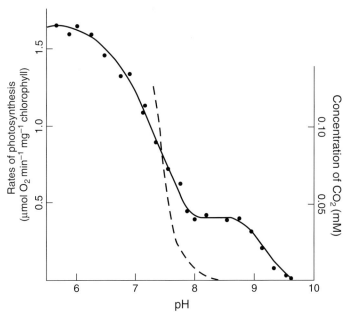

Figure 7.12   Rate of photosynthesis in the seagrass *Cymodocea nodosa*, and CO_2 concentrations, as a function of pH in a closed system such as depicted in Figure 7.4. See the text for interpretations of the results. Reprinted from: Beer S, Waisel Y, 1979. Some photosynthetic carbon fixation properties in seagrasses. Aquatic Botany 7: 129–138. Copyright (2013), with permission from Elsevier.

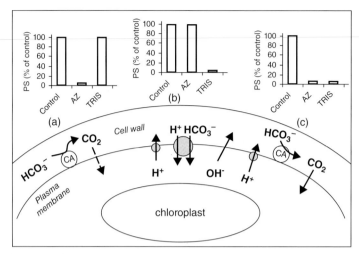

Figure 7.13   Mechanisms of Ci utilisation in seagrasses. The three most common ways of Ci utilisation (a–c) are indicated by the seagrasses' photosynthetic responses to the carbonic anhydrease (CA) inhibitor acetazolamide (AZ) and the organic buffer TRIS (as shown in the staple diagrams). Since there is a net alkalinisation of the bulk seawater pH during photosynthesis, OH- efflux is envisioned to occur at sites along the cell membrane that are spatially separated from those of H^+ efflux (for mechanisms b and c). Grey circles indicate active transport. Reprinted from: Beer S, Björk M, Hellblom F, Axelsson L, 2002. Inorganic carbon utilisation in marine angiosperms (seagrasses). Functional Plant Biology 29: 349–354. Copyright (2013), with permission from CSIRO Publishing, (Available at: www.publish.csiro.au/nid/102/paper/PP01185.htm).

from the figure, additional CO_2 in the presence of saturating HCO_3^- enhances photosynthetic rates up to several times. In other words, the photosynthetic potential in this seagrass was thought to be high, but was limited by its HCO_3^--utilisation system. This seemed, over the next couple of decades, to be true for most seagrasses studied, and while both seagrasses and macroalgae utilised HCO_3^-, the principal difference between the two plant groups was that the latter's photosynthesis saturated at the present-day seawater-Ci composition while the former's did not (see, e.g. the forthcoming Figure 8.15 in Part III of this book).

The dogma that macroalgae are saturated by today's seawater-Ci composition while seagrasses are not (e.g. by the Beer and Koch, 1996, paper) was challenged in a paper by Schwarz and co-workers, 2000: Using underwater PAM fluorometry (see Section 8.5), they showed

that two tropical seagrasses growing in shallow waters did not increase their photosynthetic rates significantly when Ci was added to the ambient seawater; thus the latter was saturating for their photosynthesis. This break of dogma can been ascribed to the way we (mis)treat especially seagrasses when doing laboratory measurements: Often seagrasses are collected from their natural environment by cutting the leaves off from the roots and rhizomes, and the latter are then further cut down to fit small (often only a few ml) O_2-measurement chambers. Bearing in mind that seagrasses are higher plants, it is not surprising that such a treatment can injure them to a degree where they do not perform optimally, and this may be the reason that their photosynthetic rates when measured this way are below those of naturally growing plants. (More about this in Section 8.5; see, e.g. Figure 8.17.)

7.4.2 Mechanisms of HCO_3^- use

Photosynthesis of many seagrasses is sensitive to the extracellularly acting carbonic anhydrase-(CA)inhibitor acetazolamide (AZ). This indicates that, just like in many macroalgae (see the previous section), CA-catalysed conversion of HCO_3^- to CO_2 within their diffusion boundary layer (DBL) precedes the uptake of the latter Ci form (see "a" in Figure 7.13). This system is operable, e.g. in the Australian species *Posidonia australis* and, possibly, also in *Posidonia oceanic a* which is endemic to the Mediterranean Sea. Like for macroalgae, this system would work best if the DBL in the vicinity of Ci utilisation were acidified, and so it is conceivable that a pH lower than that of seawater would be present in the DBL close to the Ci-utilisation site so as to promote a high concentration of CO_2 in equilibrium with HCO_3^-, but, unlike for some macroalgae, pH has never been measured within the DBL of seagrass leaves.

There exist also other HCO_3^- utilisation mechanisms in seagrasses, some of which were detected rather recently. One interesting different way of using HCO_3^- was spotted by mistake when using the buffer tris(hydroxymethyl)-aminomethane (TRIS) for keeping a constant pH in photosynthetic experiments with eelgrass (*Zostera marina*); it was found in Lennart Axelsson's laboratory that the addition of TRIS buffer *per se*, at the same pH as seawater (8.1–8.2), reduced photosynthetic rates considerably for this plant. The logical interpretation for this was that the buffer, the purpose of which is to adjust protons (H^+), did so by absorbing H^+ that were apparently extruded from the leaves, and that this somehow interfered with their photosynthetic utilisation of Ci. The mechanism involved in the Ci-acquisition system of *Zostera* as illustrated in "b" of Figure 7.13 can be seen thus: Protons are extruded by an active

H^+ pump somewhere along the leaf, possibly within a micro-zone of a cell membrane (while OH^- may be extruded somewhere else if water is split into H^+ and OH^- by the 'pump'). It is then envisioned that the surplus of H^+ is transported back into the cells, accompanied by HCO_3^-. Such a symport would indeed require a transporter to be present in the cell membrane, but this transporter has not been identified yet. Another possibility is that HCO_3^- were transported into the cells at a site where OH- was extruded in a process called antiport; it is virtually impossible to experimentally differentiate such a system from the suggested symport one.

A third way of utilising HCO_3^- in seagrasses is a combination of extracellularly acting CA and H^+ pumping. This mechanism is illustrated in "c" of Figure 7.13, and can be identified by photosynthesis being sensitive to both AZ and TRIS. If so, H^+ pumps generate a low pH within the DBL close to the site of Ci utilisation, which is conducive to a high concentration of CO_2 in equilibrium with HCO_3^-, and CA catalyses the efficient conversion of HCO_3^- to CO_2 as the latter diffuses inwards, where it is used for photosynthesis. This type of Ci utilisation is found in several tropical seagrasses such as the common Indo-Pacific species *Halophila stipulacea* (depicted in Figure 2.6a).

A few characteristics can be mentioned regarding seagrasses that may be relevant in understanding their photosynthesis: First, the chloroplasts are located to the epidermis of the leaves, i.e. to the cells that are outermost in their cross section and, thus, closest to the surrounding seawater. This ensures that the chloroplasts therein receive maximum light and that the transport or diffusion path for Ci, be it CO_2 or HCO_3^-, is as short as possible. Secondly, just as the lacunae in the leaves are necessary for transporting O_2 downwards to the underground rhizome and root system (see Section 2.5), especially during the day when O_2 is produced photosynthetically in the leaves, they may also be important in transporting CO_2 from the (usually CO_2-rich) sediment towards the leaves. While this is a logical theoretical possibility, especially during potential Ci-restricting seawater conditions (e.g. at normal Ci and such high pH conditions in a bay where both CO_2 and HCO_3^- are limiting), its quantitative importance needs yet to be established. Thirdly, like for macroalgae (see the preceding section), ^{14}C-pulse-chase experiments have shown that seagrasses are C_3 plants. However, their photosynthesis is not much affected by O_2 at normal, air-equilibrated, Ci and O_2 levels, and so most species possess a CO_2 concentrating system (CCM); if so, this CCM is, like in macroalgae, based on HCO_3^- utilisation rather than on C_4 or CAM metabolism. However, it was recently shown that *Zostera marina* and *Ruppia maritima* featured increased rates of photorespiration under conditions of increasing pH and Ci-limitations (Figure 7.14); we see such environmental conditions in many both tropical and temperate seagrass meadows.

7.5 Calcification and photosynthesis

The deposition of calcium carbonate ($CaCO_3$) as either calcite or aragonite in marine organisms (see Box 2.3) can occur within the cells, but for macroalgae it usually occurs outside of the cell membranes, i.e. in the cell walls or other intercellular spaces. The calcification (i.e. $CaCO_3$ formation) can sometimes continue in darkness, but is normally greatly stimulated in light and follows the rate of photosynthesis. During photosynthesis, the uptake of CO_2

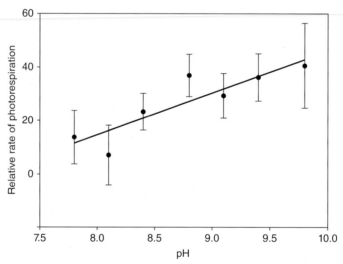

Figure 7.14   Relative rates of photorespiration of the seagrass *Zostera marina* as a function of pH (using rising pH values as a proxy for increasing Ci depletion). Photosynthesis was measured in a closed system where *Ulva* had depleted the Ci content of the seawater so that pH increased (according to Figure 3.6) so as to simulate field conditions (as illustrated also in Figure 11.1). Adapted after: Buapet P, Rasmusson LM, Gullström M, Björk M, 2013. Photorespiration and carbon limitation determine productivity in a temperate seagrass system. PLoS ONE 8(12): e83804.doi:10.1371/journal.pone.0083804. Copyright (2013) by the authors.

will lower the total amount of dissolved inorganic carbon (Ci) and, thus, increase the pH in the seawater surrounding the cells, thereby increasing the saturation state of $CaCO_3$. This, in turn, favours calcification (see 7.5 and 7.6 below). Conversely, it has been suggested that calcification might enhance the photosynthetic rate by increasing the rate of conversion of HCO_3^- to CO_2 by lowering the pH. Respiration will reduce calcification rates when released CO_2 increases Ci and/but lowers intercellular pH.

$$CO_2 + H_2O \leftrightarrow H_2CO_3 \leftrightarrow HCO_3^- + H^+$$
$$\leftrightarrow CO_3^{2-} + 2H^+ \qquad (7.5)$$

$$Ca^{2+} + 2HCO_3^- \leftrightarrow CaCO_3 + H_2O + CO_2$$
$$(7.6)$$

The saturation of seawater with respect to $CaCO_3$ is given by the saturation constant Ω:

$$\Omega = \frac{\left[Ca^{2+}\right]\left[CO_3^{2-}\right]}{[K'_{sp}]} \qquad (7.7)$$

where K'_{sp} is the stoichiometric solubility product of $CaCO_3$ (aragonite or calcite). Since the Ca^{2+} concentration ($[Ca^{2+}]$ in 7.7) is essentially constant in seawater, $CaCO_3$ formation is thus determined by $[CO_3^{2-}]$, the occurrence of which is dependent on pH. Thus, as pH rises due to CO_2 depletion, $[CO_3^{2-}]$ and Ω will also increase and calcification will be enhanced. By the same principle, ocean acidification diminishes calcification because its cause, the elevated CO_2 concentration, will cause a drop in pH and a shift in the Ci equilibria away from CO_3^{2-}, thereby decreasing Ω.

The calcification process in macroalgae is, unlike in corals, a rather straightforward matter of precipitation of $CaCO_3$ crystals within or outside the cell walls. See further about the interplay between macrophyte photosynthesis and calcification in Section 11.3.

Among the microalgae, coccolithophores produce $CaCO_3$ (mostly in the form of calcite) scales/plates in coccolith vesicles (CV) inside the cells, and these plates (coccoliths) are then extruded to the outside of the cell. Such coccolithophores (see Figures 7.15 and 2.5a) can form huge surface blooms in the oceans with anything up to 1 tonne of calcite km^{-2} in patches where they occur. Eventually, those scales sink down to the bottom of the sea, where they can accumulate into thick layers (see Section 2.2, Box 2.3 and Figure 2.5b). So the coccoliths are assembled within the CV where calcium carbonate (as calcite) develops from a ring of crystals, which is then built up into a complex 3-dimensional structure before extrusion to the surface (see Figure 2.5a). Coccolith formation involves an 88.6 KDa polysaccharide which also forms a 300 nm coating on the completed coccolith.

Although there have been a number of models put forward to describe calcification in coccolithophores, it is generally believed that HCO_3^- is transported into the coccolith vesicle, where it reacts with Ca^{2+} ions to form $CaCO_3$ and CO_2 (Figure 7.15). It is also believed that CO_2 is the source of Ci for photosynthesis, though there is considerable debate about whether that CO_2 is derived from external CO_2 (transported actively across the plasma membrane and/or chloroplast membranes), converted by external CA from external HCO_3^- or indeed derived from CO_2 released during the synthesis of calcite. It is clear though that whereas some CO_2 released from calcification may be used in photosynthesis, this is not an obligate mechanistic relationship.

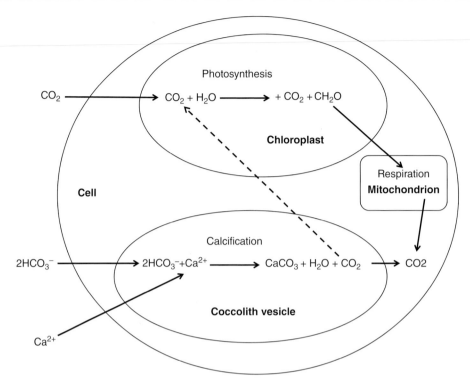

Figure 7.15   A model of the use of inorganic carbon for photosynthesis and calcification in coccolithophores. Calcification requires the active transport of HCO_3^- and Ca^{2+} ions into the coccolith vesicle. Here, they combine to form $CaCO_3$ (calcite) and CO_2 is released. In calcifying coccolithophorids (and most are), CO_2 is the dominant source of Ci for photosynthesis. While most CO_2 is supplied from outside the cell, in theory some can come from the CO_2 released in calcification, though it is clear from recent experimental data that this is not an obligate relationship (i.e. coccolithophores do not rely on internal generation of CO_2 to supply photosynthesis) and this does not act as a CCM as had previously been suggested. Drawing by John Beardall.

Summary notes of Part II

- **Photosynthetic pigments** 'capture' photons (of light). The main ones are **chlorophylls, carotenoids** and, in cyanobacteria and red algae, the **phycobilins** phycocyanin and phycoerythrin. These pigments are associated with the **thylakoid membranes** (as parts of the chloroplast **grana**) of the chloroplasts (in eukaryotic plants; the equivalent of thylakoids are invaginations of the inner cell membrane in cyanobacteria), where they are either 'anchored' or, in the case of phycobilins, protruding from the membranes in structures called phycobilisomes.

- **Photon energy** captured by most photosynthetic pigments of the thylakoid membrane is channelled (via resonance transfer) towards the **reaction centres'** 'special pair' chlorophyll molecules **P680** (for **photosystem II**, PSII) and **P700** (for **photosystem I**, PSI). P680 and P700 can, upon excitation, transfer electrons to an electron acceptor (pheophytin in PSII and 'A' in PSI), whereupon they temporarily become ionised. Thus, $P680^+$ is a strong oxidant that can 'draw' electrons from water, and $P700^+$ can draw electrons from plastocyanin in PSII. Ferredoxin is ultimately reduced in PSI, and can in turn reduce $NADP^+$ to **NADPH**. The 'linear' movement of electrons through the photosystems is called the **Z-scheme** of electron transport. Alternative routes of electrons are also possible (cyclic electron flow or the Mehler reaction as part of a water–water cycle).

- **ATP** is generated in photosynthesis (**photophosphorylation**) through the accumulation of protons (H^+) in the **thylakoid lumen**, which gives rise to a **proton-motive force** (PMF). These H^+ are released to the lumen as water is 'split' to electrons, O_2 and H^+ at the lumen side of PSII. Another source of H^+ in the lumen is generated by the **plastoquinone 'pump'**. ATP is formed as the PMF 'forces' the H^+ through **ATP synthase**, which is also located at the thylakoid membrane.

- The NADPH and ATP generated by the light reactions of photosynthesis are used in the **stroma** of the chloroplast (or the inner intermembrane spaces of cyanobacteria) to **reduce CO_2 to carbohydrates** (**sugars**). This is done via a series of reactions comprising the **Calvin cycle**, which 'starts' with the fixation of CO_2 (= **carboxylation**) to the 5-carbon compound ribulose bisphosphate(RuBP) as catalysed by the enzyme **ribulose-bisphosphate carboxylase/oxygenase** (**Rubisco**). The sugars formed in the Calvin cycle are triose phosphates (TP), and 1/6 of the TPs are exported from the chloroplast to form other sugars, while 5/6 of the TPs are used for regenerating RuBP.

- Rubisco can also act as an **oxygenase** (i.e. 'fix' O_2 onto RuBP), which is the starting point of **photorespiration**. Photorespiratory losses of previously fixed CO_2 may account for up to 25% of the photosynthetic gain in **C_3 plants** (= those plants where the Calvin cycle is the only way of carboxylation). In the terrestrial **C_4** and **crassulacean acid metabolism** (**CAM**) plants, CO_2 is concentrated to the vicinity of Rubisco by an additional, auxiliary, CO_2 'pump', which thus forms a **CO_2 concentrating mechanism** (**CCM**). CCMs in marine plants are not commonly based on C_4 or CAM metabolism, but on the use of HCO_3^- from seawater.

- The **quantum yield** of photosynthesis (Φ or Y) is defined as the quantity of photosynthesis performed per photons absorbed by the photosynthetic pigments. Thus, the maximal quantum yield (assuming no energy losses in photosynthesis) is 0.5 for electron transport (two photons needed to 'push' an electron through the Z-scheme) and 0.125 for CO_2 fixation or O_2 evolution (it takes 4 electrons to generate the 2 molecules of NADPH needed to reduce 1 molecule of CO_2, and 1 molecule of O_2 is generated as the 2 water molecules are 'split' to generate the 4 electrons). The quantum yield is approximately the same for photons of all energy levels (from 400 nm, or blue, to 700 nm, or red ones) since surplus energy above that of red photons are dissipated as heat during the de-excitation of 'excited' photosynthetic-pigment molecules.

- **Photosynthesis of marine plants** differs from that of terrestrial plants mainly by the way that they **acquire inorganic carbon** (**Ci**) from seawater. **Bicarbonate** ions (**HCO_3^-**) is the major Ci form used, and it may be either **converted to CO_2 externally** to the cell membrane (in the periplasm, as catalysed by **carbonic anhydrase**), or may be **transported into** the photosynthesising cells.

- Because of the **low diffusivity of CO_2 in water**, today's dissolved CO_2 concentration (*ca.* 13 μM) is a limiting resource for marine plant photosynthesis. Because of this, **CCMs are present** in almost all marine plants. These CCMs are based mainly on various forms of **HCO_3^- utilisation**, and may raise the intra-chloroplast (or, in cyanobacteria, intracellular or intra-carboxysome) CO_2 to several-fold that of seawater. Thus, Rubisco is in effect often saturated by CO_2, and **photorespiration is therefore often absent or limited** in marine plants.

A variety of macroalgae in the Mediterranean. Photo by Katrin Österlund.

Part III

Quantitative Measurements, and Ecological Aspects, of Marine Photosynthesis

Introduction to Part III

We have, in Part II of this book, treated photosynthesis mainly at its mechanistic level, and have in Chapter 7 focused on how marine plants acquire inorganic carbon (Ci) for this process. However, as the title of the book promises, we also have an obligation to treat marine photosynthesis in relation to the marine environment and changes therein, such as are encountered, e.g. along a depth gradient or dielly in the intertidal, as well as pertaining to interactions with other processes and other organisms. Regarding other intracellular processes, importantly, photosynthesis interacts with respiration in terms of CO_2 and O_2 exchange, and so when measuring **apparent** (or what is often called **net**) photosynthetic rates as rates of exchange of these gases (be it *per se* or for elucidating photosynthetic pathways and mechanisms), it is not always apparent what the contributions of respiration are. In principle, one can say that photosynthesis is a function of irradiance while respiration occurs both in the light and in the dark. Therefore, **true photosynthesis** (also called **gross** photosynthesis) can be derived if we correct the gas-exchange rates in the light (**apparent photosynthesis**) for the rate of respiration measured in the dark. This, however, will yield an approximation of photosynthetic rates only since a) other types of processes that consume O_2 (e.g. the Mehler reaction) also occur in the light, b) rates of mitochondrial respiration (sometimes called dark respiration) can be affected by the irradiance (see Box 8.1), and c) there may be photorespiration present (although the CO_2 concentrating mechanism, CCM, of many marine plants represses this process). Often, however, this approximation is sufficient for the scientific questions asked. There is an increasingly popular way of measuring true photosynthetic

Photosynthesis in the Marine Environment, First Edition. Sven Beer, Mats Björk and John Beardall.
© 2014 John Wiley & Sons, Ltd. Published 2014 by John Wiley & Sons, Ltd.
Companion Website: www.wiley.com/go/beer/photosynthesis

rates where respiration does not affect the measurement, i.e. the pulse-amplitude modulated (PAM) fluorometric method in which photosynthetic electron flow can be derived from quantum yield measurements, and this will be discussed at length in the forthcoming chapter.

After outlining some common methods used in photosynthetic research of marine plants in the first chapter of this part of the book, and defining more precisely apparent and true photosynthesis as a function of irradiance, we will in subsequent chapters describe how photosynthesis responds to various environmental conditions in the sea, and illustrate cases where different photosynthetic strategies can both cause acclimation and adaptations[1] of plants to specific environments and indeed how they may influence those environments and, thereby, also influence other organisms living within them. Before doing so, however, a caution (or apology) is needed: While we begin to know how marine plants photosynthesise, we are much more lacking in knowledge about how various photosynthetic mechanisms can confer advantages in different environments. We are also, in this year of 2013, very far from understanding how various strategies of using Ci photosynthetically can influence other organisms in a symbiotic or competitive way. This is thus a rather novel area of marine science of which we know only little, and it will largely become the challenge for younger, upcoming, scientists in the field of marine photosynthesis and ecophysiology to further it. Since we, the authors of this book, are aging out of science (but still remain good-hearted), we will try to point out where progress can be made by others, and possibly how.

[1] 'Acclimation' is here used as a short-term, plastic, response to change in (an) environmental factor(s) while 'adaptation' refers to a longer-term selection process, ultimately resulting in a genetically different ecotype).

Chapter 8
Quantitative measurements

Photosynthesis (or, more correctly, the rate of photosynthesis) in terrestrial environments is measured as the exchange of O_2 or, more commonly, CO_2 or, often today, by fluorometry. In marine environments, a long-standing approach has been to 'spike' CO_2 with the radioactive isotope ^{14}C, which can be easily supplied in the form of $H^{14}CO_3^-$ (e.g. as $NaH^{14}CO_3$). Thus, with the knowledge of the specific activity of ^{14}C (radioactivity per, e.g. mol of C), the Danish marine botanist Einer Steemann Nielsen in his book Marine Photosynthesis (1975) described ways to measure the photosynthetic productivity of marine plants. In principle, the inorganic carbon (Ci) source containing ^{14}C is added to benthic plants or phytoplankton assemblages within enclosures for certain amounts of time, after which the plants are collected (with plankton nets or by filtration in the case of phytoplankton) and killed in an acid, the latter also releasing non-assimilated ^{14}C as $^{14}CO_2$. The radioactivity within the organic matter of the plants is then measured, and photosynthetic rates calculated by dividing this radioactivity by the specific activity of ^{14}C. With today's knowledge of isotope chemistry, one can also correct for the discrimination by Rubisco against the slightly larger ^{14}C compared to ^{12}C. Two significant problems with this method had (and have, where still used) to be taken into consideration when calculating photosynthetic rates from the ^{14}C-activity measurements: First, a significant portion of $^{14}CO_2$ was incorporated in the dark, and such non-photosynthetic CO_2 fixation has to be taken into account. Secondly, and as part of the first problem, it was not clear if the method measured true or apparent photosynthesis. Initially, no $^{14}CO_2$ is evolved by respiration. However, with time as the intracellular carbohydrate pool gets increasingly labelled with ^{14}C, the plants respire $^{14}CO_2$ at increasing rates and, so, apparent rates of photosynthesis are measured. These uncertainties, together with an increasing awareness that we should not pollute the environment with radioisotopes, and the fact that fluorometry has lately become a sensitive and reliable tool, have resulted in the $^{14}CO_2$ method not being used much today for measurements of photosynthetic rates in marine plants. (For those interested in the method, we refer to the book by Steemann Nielsen.) Therefore, we will here focus on gas-exchange measurements of O_2

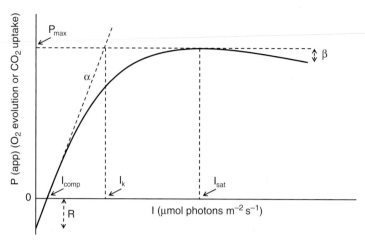

Figure 8.1 A 'typical' photosynthesis (*P*) *vs.* irradiance (*I*) curve. See text for explanations. Drawing by Sven Beer.

and CO_2, and then on the increasingly popular method of pulse-amplitude modulated (PAM) fluorometry.

Photosynthesis can of course be measured as a function of various parameters, e.g. the concentration of Ci or other nutrients or concentrations of inhibiting pollutants etc. When measured as a function of irradiance, however, then there are certain key points along the resulting **photosynthesis-irradiance**[1] (or ***P–I*) curves** that bear a nomenclature to which most investigators adhere, and this is illustrated in Figure 8.1. Here, **apparent photosynthesis**, P(app), is measured in a closed system as changes in the rate of either O_2 evolution or CO_2 uptake (calculated from concentration changes in the surrounding air (for terrestrial plants) or water (for submersed plants) per time). In darkness (i.e. zero irradiance), the rate of respiration (R), as O_2 uptake or CO_2 evolution, is obtained. As irradiance increases, photosynthesis increasingly releases O_2 and con-

sumes CO_2 until a point is reached where no net change in these gases can be noticed, i.e. the rate of photosynthesis equals that of respiration and the resulting net gas-exchange rate appears as zero. The irradiance at which this occurs is called the **light-compensation point** (I_{comp} on the x-axis). As irradiance is further increased, photosynthesis increasingly evolves O_2 (and assimilates CO_2) so that P(app) increases to positive values until **saturation** is reached at the rate P_{max}. The corresponding irradiance at which P_{max} is reached is accordingly termed I_{sat}. Also displayed in the figure are the **initial slope of photosynthesis**, α, or what is sometimes taken as the 'maximal quantum yield' (see however under 'Alpha, "uses and misuses"' in Section 8.5), and the component of **photoinhibition** at high irradiances, β. Alpha is obtained as a line tangential to the portion of the $P–I$ curve where P(app), or also P(true), see Figure 8.2, increases linearly with irradiance, i.e. at low irradiances. (See more about α later, and about photoinhibition too.) Since it is hard to derive an exact I_{sat} from the $P–I$ curve, there is another parameter that is easier to determine: I_k. The latter is defined as the crossing point of the extended line of α with that of the

---

[1] Photosynthesis *vs.* irradiance (*P–I*) curves are sometimes, or even most often, termed photosynthesis *vs.* 'energy flux' (*P–E*) curves. We will, however, here adhere to the term *P–I*.

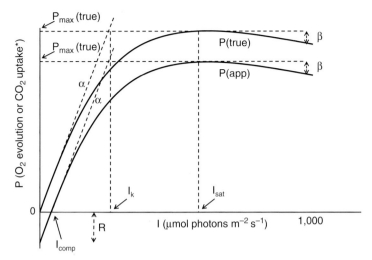

Figure 8.2   **Apparent**, P(app) **and true**, P(true), **rates of photosynthesis** (P) as a function of irradiance (I), or so-called P–I curves. On the P(app) curve, the rate of respiration (R) is obtained in darkness. Rates of true photosynthesis can then be derived from the P(app) curve by correcting for R, and this yields the P(true) curve. See the text for further explanations; see also the text pertaining to Figure 8.1 for definitions of the various abbreviated terms, and see the following box on effects of light on respiration. Drawing by Sven Beer.

extended line of P_{max}, and is termed the '**onset of light saturation**'. A P–I curve is generated by measuring photosynthetic rates at various irradiances, and then either plotting the results or inserting them into the equations of Jassby and Platt (1976) or Platt, Gallegos and Harrison (1980; the latter contains the addition of a photoinhibition term, β). The various parameters (α, P_{max}, etc.) can then either be derived graphically or from the equations.

8.1   Gas exchange

When measuring photosynthetic rates as gas (O_2 or CO_2) exchange, it is always apparent rates that are measured (except for sophisticated laboratory measurements using mass spectrometry, which we will not discuss here). This means that part of the exchange of gases stem from respiration, and so if true rates of photosynthesis are sought, then the apparent rates have to be corrected for rates of respi-

ration. If we decide to live with the approximation that **true** (or gross) **photosynthesis** equals **apparent** (or net) **photosynthesis** measured as gas exchange in the light **corrected for the O_2 taken up or CO_2 released by respiration** as measured during darkness (see however Box 8.1), then Figure 8.2 illustrates the relationship between the two. Such measurements can be carried out in the laboratory, e.g. as illustrated in Figure 7.4 (in the case of O_2), or in the field as will be described later. Thus, rates of true photosynthesis, P(true) in the figure, are derived by correcting the measured apparent photosynthesis with the rate of respiration in the dark (R). Note that R is strictly speaking negative for changes in O_2 (O_2 is taken up from the seawater medium in respiration) and positive for changes in CO_2 (CO_2 is released by respiration). Therefore, in order not to mix up pluses and minuses, we say that P(true) equals P(app) **corrected for** R, yielding always higher rates of photosynthesis for P(true). Note that only P is different between the two curves

Box 8.1   Effects of light on 'dark' respiration

For calculations of rates of true photosynthesis it is convenient to assume that mitochondrial (or 'dark') respiration is constant and does not change dielly. However, it is known that respiration can be quite substantially down-regulated in the light in many plants and enhanced following high rates of photosynthesis in others, and that there is thus a significant variation in the respiration rate during the day and, interestingly, also the night.

Several studies performed on terrestrial plants led to the conclusion that light affects mitochondrial respiration negatively. When the light period starts in the morning, the activity of some key enzymes of the respiratory chain become inhibited or down-regulated. As a possible result, photosynthates accumulate increasingly during the light period, and those are then available just after darkness when there is no longer any down-regulation of respiration. This may accordingly result in a short increase in respiration just after the end of the light period (not to be confused with a post-illumination burst of CO_2 caused by photorespiration). Alternatively, the down-regulation of mitochondrial respiration in the light may be modulated by an increased potential for respiration as photosynthates accumulate. Whichever the case, the net result is a rate of mitochondrial respiration that is affected by the light, but its true value may be hard to measure since photosynthesis always 'masks' respiration during the day.

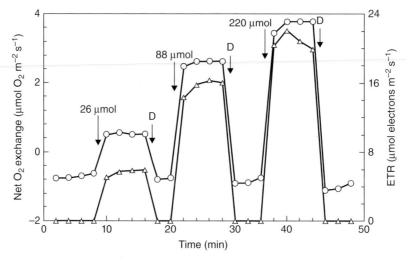

Figure 8.3   Time course of net O_2 exchange (circles) **and photosynthetic electron transport** (ETR, triangles) **rates** for *Ulva lactuca* at three different irradiances (in μmol photons m^{-2} s^{-1}) and intermittent periods of darkness (D): While ETR is zero at darkness, the O_2 exchange trace shows increasing rates of dark respiration (more negative rates) as the preceding irradiance increases. Reprinted from: Beer S, Larsson C, Poryan O, Axelsson L, 2000, Photosynthetic rates of Ulva (Chlorophyta) measured by pulse amplitude modulated (PAM) fluorometry. European Journal of Phycology 35: 69–74. Copyright (2000) with permission from Taylor & Francis.

The fact that light affects rates of dark respiration in marine plants has been demonstrated for, e.g. the macroalga *Ulva lactuca* (Figure 8.3): When thalli were subjected to light, then the rates of respiration following darkness were initially higher than after a minute or two in darkness. Also, and more strikingly, those initial rates were higher in a way that directly depended on the previous irradiance. We (MB and SB) have interpreted this phenomenon as being due to the carbohydrates formed increasingly at increasing irradiances being directly available for respiration in the light; *Ulva* features only chloroplast-containing cells, and the spatial proximity of photosynthesis and respiration within all cells make this suggestion possible. If a similar initial high rate of O_2 uptake, or burst of CO_2 (a so-called 'post-illumination CO_2 burst'), were to be observed in terrestrial C_3 plants, then it would be ascribed to a continuation of photorespiration. However, since photorespiration is virtually absent in efficiently photosynthesising marine algae such as *Ulva* (largely because of their CO_2-concentrating mechanism, CCM, see Section 7.3), it is here ascribed to the effect of light on mitochondrial respiration and this phenomenon is also well described in microalgae. Similar experiments on the seagrass *Posidonia oceanica* showed that mitochondrial respiration fluctuated also during the night-time, but the reason for this is still far from being understood.

In conclusion, rates of mitochondrial respiration vary dielly, and/but since they are hard to measure (photosynthesis 'masks' respiration in the light), for the lack of anything better we will presently often have to live with the approximation that the rate of true photosynthesis as measured by gas exchange is that of apparent photosynthesis corrected for dark respiration (measured at some notional time, but usually at steady state after 10–15 min in darkness).

in Figure 8.2 (including P_{max}); I_{sat}, α and β, as well as I_k, remain the same for the two curves. Depending on the way photosynthesis is measured, the rates on the y-axis (*) can be expressed as μmol O_2 m^{-2} s^{-1} or μmol CO_2 m^{-2} s^{-1} (and electron-transport rates (derived from PAM fluorometry) as a measure of true photosynthesis is also expressed on a m^{-2} s^{-1} basis, see Section 8.3). The reason for expressing photosynthetic rates on a per m^2 per second basis is because light is measured on such a basis (μmol photons m^{-2} s^{-1}, see Section 3.1).

8.2 How to measure gas exchange

In the previous section, we noted that photosynthesis can be measured as gas exchange of either O_2 or CO_2. 'Exchange' implies the exchange of O_2 or CO_2 between the plant and its environment, and is measured in the plants' enclosed surroundings.

Oxygen is easily measured in aqueous media by rather inexpensive **O_2 electrodes**. Most commonly used are those of the **Clark-type** (named after Leland Clark, who designed them for measurements of blood-O_2 levels). The principle by which Clark-type electrodes operate is that O_2 equilibrates across a membrane and is, on its inward side, reduced to H_2O_2 by electrons carried along a polarising voltage potential of $\sim$0.7 V within an electrolyte solution. At this voltage, and given the materials of the anode (silver) and cathode (platinum), the electrodes become highly specific for O_2 (but are notoriously sensitive also to sulfides), the concentration of which is proportional to the current generated as O_2 is reduced. Two features need to be considered when using such electrodes: 1) Diffusion, including that of O_2, is a rather slow process, and the medium close to the membrane has to be stirred vigorously in order to generate a flux across it that corresponds to the external O_2 concentration, and 2) Clark-type O_2 electrodes are very temperature sensitive such that temperature needs to be kept constant within some 0.1 °C. (This sensitivity is much higher than the sensitivity of O_2 dissolution at different temperatures.) If those criteria are met, then a typical O_2-electrode system for laboratory measurements is that described in Figure 7.4. While Clark-type O_2 electrodes are sold by several manufacturers, one can also buy setups where they are complemented with water-jacketed (for keeping constant temperature) incubation chambers into which plants or plant parts can be inserted and light sources are provided (including those made by Hansatech, UK); in the latter, only a constant-temperature water bath needs to be fitted in order to circulate water of a constant temperature through the water jacket such as illustrated in Figure 7.4 (although that specific water jacket was homemade). While the generally small volume (a few ml) of the incubation chambers usually provided for O_2 electrode setups makes them suitable for phytoplankton and small macrophytes, most marine macrophytes are too large to be fitted into them, and cutting up such plants may not yield reliable results (see Section 8.5 under 'Measuring whole plants'). For field work, while portable, battery-driven, Clark-type O_2 electrodes may be used *in situ*, it may often be worthwhile to withdraw water samples from the incubation chambers periodically and measure their O_2 concentration on the beach or in the laboratory.

Another type of sensor available for O_2 measurements is of the **optode** type. The principle of their operation is that fluorescence generated by specific compounds is quenched by O_2. Thus, if these compounds are present in a matrix exposed to O_2 then, for instance, a red fluorescence signal generated by illuminating with blue light will be quenched depending on O_2 concentrations. These sensors are becoming increasingly popular as compared to the electrochemically based Clark-type electrodes because they have no membrane or electrolyte that have to be replaced periodically, and they do not depend on stirring in their vicinity since they do not use up any O_2 themselves. They are also much less temperature sensitive than electrodes. Also, field optodes exist that save data internally, and they can be placed in natural settings for extended time periods. On the other hand, optodes are generally less sensitive to changes in O_2 concentrations than Clark-type electrodes, especially at high concentrations (their response *vs.* O_2 is non-linear).

A 'typical', or at least 'principal', experimental setup in which photosynthetic rates can be measured as O_2 exchange is outlined in part "A" of Figure 8.4: Plants are enclosed in seawater, either in the laboratory (usually under artificial irradiation) or in the field (under natural light, see also in Section 12.3 for *in situ* incubations), and changes in O_2 concentrations with time (increase if apparent photosynthesis is positive and decrease if it is negative) are measured by an O_2 electrode connected to an O_2 meter, and are recorded either by a chart recorder or, today, by a data-acquisition programme in a computer. The changes in O_2 concentration are then recalculated to amounts of O_2 by multiplying by the volume of seawater used, and the results can be given as, e.g. μmol O_2 m^{-2} (of plant surface) s^{-1} or, if phytoplankton are measured, per mg chlorophyll (see standard handbooks of oceanography for chlorophyll extractions and determinations) or some other plant parameter. For calibration of the O_2 electrode, there are tables and, today, programmes that calculate air-equilibrated O_2 concentrations as a function of temperature and salinity (e.g. SWWIN.exe, the availability of which can be sought from the authors; see also at the end of Section 3.3 for Internet sites where similar programmes can be found). For example, at 20 °C and a salinity of 35, the O_2 concentration in seawater equilibrated with the atmosphere

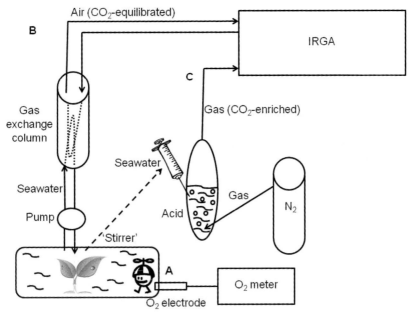

B

Air (CO_2-equilibrated)

IRGA

C

Gas exchange column

Gas (CO_2-enriched)

Seawater

Seawater

Gas

N_2

Pump

Acid

'Stirrer'

A

O_2 meter

O_2 electrode

Figure 8.4   Principles of gas-exchange measurements of O_2 (A) and CO_2 or Ci (B and C, respectively). The full arrows represent tubes through which seawater, air and gas are flowing (as indicated), and the little figure is a stirring device (e.g. the bar of a magnetic stirrer). See the above and below text for details. Drawing by Sven Beer.

containing 21% O_2 is 230 μM (i.e. μmol l^{-1}; multiplying by the number of litres (or parts thereof) of incubation medium will give the amount of O_2 in μmol).

There is a classical way to measure O_2 concentrations chemically by titration called **the Winkler method** that was developed almost 130 years ago by the Hungarian chemist Lajos Winkler. Since the dissolved O_2 can be fixed within the so-called Winkler bottles, titrations in order to find the original O_2 concentrations prior to fixation can be performed later in the laboratory. While the Winkler method may be old, it is very precise, and is used today as the 'gold standard' when, e.g. calibrating Clark-type O_2 electrodes and optodes. (See general methods texts in oceanography for details of the Winkler method.) Finally, O_2 concentrations can also be measured using micro-O_2 electrodes. They are normally Clark-type electrodes, and are called 'micro' since their tips are

very thin, sometimes less than a few micrometres (μm). This makes it possible to do very precise (and with a quick response time) measurements of dissolved O_2 in tissues, cell walls and within diffusion boundary layers; the latter has been done extensively for photosynthetic measurements in corals. The micro-O_2 electrodes need to be held at a precisely fixed distance from the photosynthesising organism or, in the case of intracellular measurements, organelles, and their use therefore requires expertise in micro-manipulations.

Carbon dioxide is measured more easily and, certainly if compared with O_2 more accurately, in air than in water. (The latter is partly because the background concentration in air, *ca.* 0.04%, is much lower than the 21% O_2, and also because the measurement by IRGA is very sensitive yet generates accurate and stable readings.) The most common way of measuring CO_2 in air is by **infrared gas analysis**

(IRGA, which is also the abbreviation of the instrument used, an infrared gas analyser). The principle underlying the method is that specific wavelengths of infrared light are absorbed specifically by CO_2. Thus, a gas is passed through the space between an infrared source and an analyser; the higher the concentration of CO_2 in the gas, the more will the infrared light be absorbed by it. Infrared gas analysis is a highly sensitive method, registering differences at ppm levels, and field instruments have been devised (e.g. by Li-Cor, USA, and ADC, UK). Because of their measurement principle, these instruments can only be used for gaseous measurements of CO_2 and are, thus, designed for measurements of terrestrial photosynthesis. However, they can be adapted for measurements of aquatic, including marine, photosynthesis. This can be done in two ways: In the first way (part "B" in Figure 8.4), water from an incubation chamber is led through a gas-exchange column in which dissolved and gaseous CO_2 are calibrated (e.g. across the walls of Teflon tubes through which the seawater flows), and the gas is then led into an IRGA. One potential problem with this method is that a low concentration of dissolved CO_2 is in equilibrium with a much higher concentration of other Ci forms at normal seawater pH (see, e.g. Figures 3.4 and 3.6). Since these Ci equilibria are extremely sensitive to pH, it is imperative that pH is kept constant throughout an incubation so as not to alter the CO_2/Ci ratios, which could easily be changed drastically by even minute changes in pH and that, thus, would not only mask, but override changes in CO_2 concentrations caused by the plant's metabolism *per se* (e.g. increased pH when photosynthesis exceeds respiration, see Section 3.2). (Keeping a constant pH during incubations could be achieved by the use of buffers, but these may have adverse effects on the plants, see Section 7.4, and so their use is not recommended.) Bearing these restrictions in mind, there is another way in which IRGA can be used for

photosynthetic measurements of submerged plants: Small aliquots of water are withdrawn from incubation chambers with a syringe, and those can be injected into an acid solution (e.g. 2% phosphoric acid) bubbled with a CO_2-free gas on-line towards an IRGA (part "C" of Figure 8.4). In this way, all Ci will be converted to CO_2 in the acid, and the concentration of the latter is recorded by the instrument. (Such recordings appear as peaks in the otherwise CO_2-deficient N_2 stream; the peak heights are calibrated with standard Ci solutions, e.g. 2 mM $NaHCO_3$, which is close to the normal seawater Ci-concentration.) Because this technique measures Ci as a whole, changes in concentrations as a result of photosynthesis may be slow since they are measured against a high background concentration of *ca.* 2000 µM.

While open systems can be used when monitoring O_2 and CO_2 or Ci fluxes, the resulting data is harder to quantify than for closed systems, such as outlined in Figure 8.4. This is because a concentration gradient of dissolved gases is generated away from the plant, including their exchange between the seawater and the overlying atmosphere. Still, such measurements can be used for estimating gas-exchange rates of marine communities (see Section 12.3). Also, open systems avoid the, otherwise unavoidable for closed systems, fact that environmental properties such as CO_2 and O_2 concentrations within enclosures change as a result of both plant and animal (if present) metabolism, as does pH, temperature, irradiance and the light spectrum. For example, if the biota within an enclosure shows positive apparent photosynthesis, then the O_2 concentration will gradually increase therein such that photorespiration could be stimulated, which would underestimate the true gas-exchange rate had O_2 been kept at ambient levels. Conversely, Ci would decrease, and since pH increases as photosynthesis proceeds, this would also reduce the CO_2/Ci ratio, in addition to the total Ci concentration, such that photosynthesis would be impeded (especially for less-efficient HCO_3^- users). Again, in open systems these restrictions are largely (or at least partly) avoided, but the data may be hard to quantify (see, however, Section 12.3).

In summary so far, while CO_2 measurements are possible in aqueous media, they are rather complicated and involve either the measurement of gaseous CO_2 equilibrated with the CO_2 dissolved in seawater or the measurement of total Ci (both using an IRGA). In the first case, even slight changes in pH of the incubation medium may alter the CO_2/Ci ratio so as to completely mask the true rate of change in CO_2 based on photosynthesis and respiration, and in the second case the high background Ci concentration ($\sim2000\ \mu M$) often yields sluggish, non-precise, results. Because of these constraints, gas exchange in the marine environment is more often measured as changes in O_2 concentration. For this, O_2 electrodes (either Clark-type electrodes or optodes) are commercially available. If high precision is sought, then the Winkler method for chemically determining dissolved O_2 concentrations can be used. For all gas-exchange methods, the plants are usually enclosed in order for CO_2 and O_2 to accumulate or diminish so that the concentration differences with time reflect the biological activities of photosynthesis and respiration by the enclosed biota. Such enclosures have their limitations too, which should be taken into account when evaluating data emanating from them (see later in this chapter).

As a final paragraph of this section, we thought of presenting the different ways of measuring photosynthetic rates, including the forthcoming PAM-fluorometric one, as part of the Z-scheme of photosynthetic electron transport (cf. Section 5.2), as illustrated in Figure 8.5. So, the O_2 we measure is generated in the light-reactions of PSII, and CO_2 is fixed and then reduced in the Calvin cycle. (Against this stands, of course, the converse exchange of these gases by respiration.) The stoichiometric

(molar) relationship between these two parameters (also termed the **photosynthetic quotient**, **PQ**) is 1 (1 O_2/1 CO_2). This is true not only for measurements of true photosynthetic rates, but also for apparent rates since the molar ratio of O_2/CO_2 (O_2 used and CO_2 evolved) in respiration is also taken as 1, though in practice this depends on the level of reduction of the products (in photosynthesis) or substrates (in respiration). For instance, phytoplankton cells using nitrate (NO_3^-) may have PQ values above 2, whereas in ammonia-(NH_3^-) grown cells PQ values are 1–1.3. The figure also shows the division of energy between photosynthesis (P) and fluorescence (Fl) during de-excitation of chlorophyll (chl) in photosystem II (PSII) that was excited by photons (Ph) (see Section 5.1). If we disregard the little heat that is also generated, then this illustrates the truism that there is an inverse relationship between P and Fl: the more photon energy used for photosynthesis, the less is left for fluorescence and *vice versa*. This inverse relationship between photosynthetic efficiency and fluorescence yield is the basis of PAM fluorometry, which will be treated in the next section. The overall theoretical stoichiometry for photosynthesis is 1 O_2/1 CO_2/4 e^-/8 Ph (see the 'Summary of Chapters 5 and 6' in Chapter 6). Similarly, the maximal quantum yield for photosynthesis (i.e. mol photosynthetic products per mol photons absorbed by the photosynthetic pigments) is 0.125 for O_2 and CO_2 (8 photons per O_2 or CO_2) and, conveniently, 1 for electron flow through PSII (which is what PAM-fluorometry measures).

8.3 Pulse amplitude modulated (PAM) fluorometry

When chlorophyll is excited by photons, part of the energy released by its subsequent de-excitation generates fluorescence (see Box 5.3

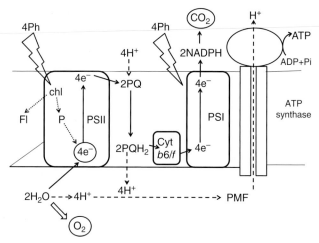

Figure 8.5   The Z-scheme and the three parameters commonly used for photosynthetic rate measurements (circled), i.e. O_2 evolved and CO_2 fixed, and electron (e^-) transport (the latter derived from measurements of fluorescence, Fl). Full arrows represent electron flow, the broken arrow proton flow and the dotted arrows on the left the division of energy between photosynthesis (P) and fluorescence (Fl) during de-excitation of chlorophyll (chl) excited by photons (Ph). See the text in the above paragraph for explanations. Except for Fl and P, all abbreviations are as in Figures 5.3 and 5.4. Drawing by Sven Beer.

and Figure 5.1). Thus, there is an inverse relationship between photosynthesis (P) and fluorescence (Fl, see Figure 8.5) or, put more exactly, between photosynthetic yield and fluorescence yield; the latter can therefore be used for calculating photosynthetic quantum yield and, from this, the photosynthetic rate – if the fluorescence can be measured! The measurement of fluorescence is not trivial since fluorescence is in essence light at a wavelength of ~700 nm emitted by the de-excitation of chlorophyll that comprises only a few per cent of the photon energy captured by this pigment, i.e. it must be detected and measured against a very high background of surrounding visible light of the same wavelength, which is, among other things, also reflected back from the leaf. The 'trick' in separating chlorophyll fluorescence from ambient light of the same wavelength lies in the principle termed **pulse-amplitude modulated (PAM)** fluorometry. For this, the PAM fluorometer emits a weak beam of modulated light in short pulses (usually

100–1000 s^{-1}) through a fibre-optic cable that is directed towards the photosynthetic organism (see Figure 8.6). The colour of this '**measuring light**' is in principle unimportant, but red (of a lower wavelength than that of the fluorescence) or blue are good since these colours are readily absorbed by chlorophyll. The PAM fluorometer is calibrated such that only the chlorophyll fluorescence stemming from this modulated measuring light is measured and all background light of the same wavelength is ignored. This measured (PAM-)fluorescence is termed **F**.

8.3.1 Quantum yields

In order to continue, we must now first agree that photosynthesis is most efficient at very low irradiances and increasingly inefficient as irradiances increase. This is most easily understood if we regard 'efficiency' as being dependent on **quantum yield**: At low **ambient**

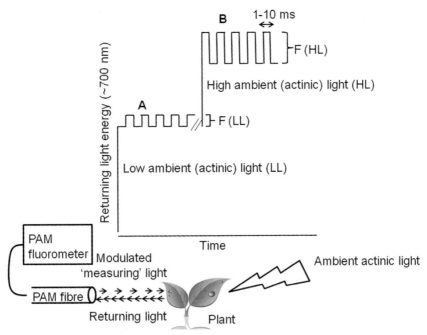

Figure 8.6   The PAM fluorometry principle: A weak beam of modulated (or pulsed), LED-based, 'measuring' light is generated by the PAM fluorometer and is directed towards a plant leaf through its optical fibre (lower part of the figure). At the same time, the plant can be irradiated by ambient actinic (causing photosynthesis) light from the sun or from another light source. The light returning to the fluorometer through the same fibre is now separated such that only the fluorescence generated by the measuring light pulses (of 1–10 ms duration) is recorded while all background light is ignored; this fluorescence is termed F, and is low at low irradiances (LL) and higher at high irradiances (HL). Drawing by Sven Beer.

irradiances (the light that causes photosynthesis is also called '**actinic**' light), almost all the photon energy conveyed through the antennae will result in electron flow through (or charge separation at) the reaction centres of photosystem II (we concentrate on photosystem II because at room temperatures fluorescence is emitted mainly from there). Another way to put this is that the chances for energy funneled through the antennae to encounter an oxidised (or 'open') reaction centre are very high. Consequently, almost all of the photons emitted by the modulated measuring light will be consumed in photosynthesis, and very little of that photon energy will be used for generating fluorescence (see part A of Figure 8.6) and the quantum yield (again, meaning photosynthetic

output per photons absorbed by the photosynthetic pigments) is high. We say that here fluorescence is largely '**quenched**' by photosynthesis, or that the efficient photosynthesis causes a high degree of **fluorescence quenching**. Consequently, the higher the ambient (or actinic) light, the less efficient is photosynthesis (quantum yields are lower), and the less likely it is for photon energy funnelled through the antennae (including those from the measuring light) to find an open reaction centre, and so the fluorescence generated by the latter light increases (see part "B" in Figure 8.6).

The fluorescence states, **F**, at low (LL) and high (HL) irradiances can also be illustrated as in Figure 8.7; the lowest fluorescence (i.e. the highest fluorescence quenching by the

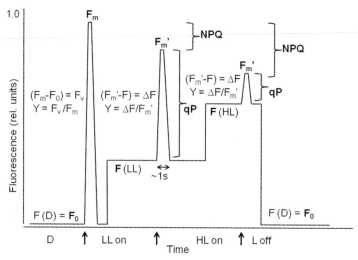

Figure 8.7   Time course of chlorophyll fluorescence emission generated by a weak beam of modulated measuring light from a PAM fluorometer: In virtual darkness (D on the *x*-axis), the fluorescence of the dark-adapted plant (F (D)) is low and is termed F_0. Upon a ~1 s saturating light flash (up arrow), fluorescence reaches a maximum (F_m) and then returns close to the previous F_0 value after the flash. The plant is now illuminated by low light (LL on) and the fluorescence increases to the value F (LL). Again, a saturating flash of light (up arrow) causes F to increase, now to the value F_m' and then back to F (LL) after it. Under higher irradiance (HL on), F increases to the value F (HL), and the saturating light pulse (up arrow) then again causes F to increase to (another) F_m' value. Following darkness (L off), fluorescence eventually decreases to the original value F_0 from where we started this journey. The quantum yield Y is maximal in virtual darkness (F_v/F_m) and then decreases as light increases ($\Delta F/F_m'$). Indicated are also values of photochemical (qP) and non-photochemical (NPQ) quenching under illumination. Drawing by Sven Beer.

'efficient' photosynthesis) is in virtual darkness, i.e. at no actinic light and only the very low (<1 µmol photons m^{-2} s^{-1}) measuring light from the PAM fluorometer reaching the plant; see state F (D) in the figure, also termed F_0 (F zero). After a stable fluorescence signal is obtained in virtual darkness or at a certain irradiance, the PAM fluorometer now performs another 'trick': strong white light, saturating for photosynthesis (>5000 µmol photons m^{-2} s^{-1}), is emitted through the fibre-optic cable for a duration of ~1 s (the time must be short so as not to cause photodamage to the plant). During this time, virtually all reaction centres become reduced (or 'closed') as they are 'busy' transferring electrons, and the chance for the photons of the modulated measuring light to transfer their energy towards

photosynthesis is close to zero. Consequently, their energy is transformed to fluorescence, the level of which is termed either F_m (maximal fluorescence, from a dark-adapted plant measured only with the weak measuring light beam) or F_m' (if measured under actinic light). In order to complicate things further, ($F_m - F_0$) is also called **variable fluorescence** (F_v) while ($F_m' - F$) is termed ΔF. ($F_m - F_0$) and ($F_m' - F$) also represent the degree of **photochemical quenching, qP**, which is highest in the dark-adapted plant (where photosynthesis is most efficient) and then decreases as irradiance increases. Now comes the beauty of PAM fluorometry: It turns out that ($F_m - F_0)/F_m$ (or F_v/F_m) and ($F_m' - F)/F_m'$ (or $\Delta F/F_m'$) equal the **quantum yield (Y)** of photosynthetic electron transfer through photosystem II.

As can be logically conceived, and as also illustrated in Figure 8.7, Y is highest for the dark-acclimated plant (approaching 1) and lowest for the high-light-acclimated one. The highest Y ever recorded for a (well-fed, well-watered and exposed to a Mozart quintet) plant in dim light is 0.84 (i.e. 83% of the photon energy captured by chlorophyll in photosystem II is converted to photosynthetic electron flow). Two additional items regarding Figure 8.7: The absolute fluorescence units are of no importance when determining quantum yields in PAM fluorometry; the way Y is calculated means it will always be a number <1.0. Secondly, while it is logical that F increases with increased irradiances, it is less apparent why F_m' decreases. This has to do with another type of quenching than qP called non-photochemical quenching (**NPQ**; there is also another way of expressing non-photochemical quenching as qN, but this requires the knowledge of F_0 during ambient irradiance and is hard to measure with, e.g. the Diving-PAM): At high irradiances, much of the photon energy is diverted to heat via, e.g. the xanthophyll cycle and other protective mechanisms (see Section 9.1). If so, then F_m' is reduced (or quenched) by those other energy-diverting pathways, and NPQ can be calculated as $(F_m - F_m')/F_m'$. As a final note, these are the **basic principles** of PAM-fluorometric measurements of Y, and other, more detailed or specialised aspects of the method can be found in, e.g. the book Chlorophyll Fluorescence: A Signature of Photosynthesis, 2004 edited by Papageorgiou and Govindjee.

8.3.2 F_v/F_m

The highest yield along an irradiance gradient is obtained in virtual darkness (i.e. at the very low irradiance generated by the measuring beam of the PAM fluorometer). Since Y is here calculated as $(F_m - F_0)/F_m$, and since $(F_m - F_0)$ is called F_v (variable fluorescence), the Y value obtained is often simply called F_v/F_m (see the left side of Figure 8.7); another appropriate name is the **maximal quantum yield** (as opposed to the **effective quantum yield** calculated as $(F_m' - F)/F_m'$ or $\Delta F/F_m'$). Since the highest F_v/F_m value may be obtained after several hours of dark-adaptation (e.g. during pre-dawn mornings), these will include, e.g. repairs (or replacement via de novo synthesis) of the D1 protein damaged by previous-day high irradiances. If this is not desired, then shorter time periods of darkness will suffice; often F_v/F_m is measured after 10–15 min of dark-adaptation. Since light is NOT a factor affecting F_v/F_m during its measurement, variations in this parameter can be used as an indicator of 'stress' affecting photosystem II that is different from irradiance. Thus, F_v/F_m has been used widely as an indicator of water stress, thermal stress, stress imposed by pollutants and herbicides, etc. Again, these 'stressful' conditions are measurable if they affect photosystem II, often at the level of the D1 protein.

8.3.3 Electron transport rates

So far, PAM fluorometry has yielded quantum yields (Y) of photosynthetic electron transport through photosystem II. Since Y can be defined as electrons transported per photons absorbed by the photosynthetic pigments associated with this photosystem, it follows that **electron transport rates** (**ETR**) can be derived if we multiply Y by the **absorbed irradiance** (I_a) of that photosystem (i.e. $I_a \cdot 0.5$, assuming a 50:50 percentage distribution of absorbed photons between the two photosystems). However, while the **incident irradiance** (I_i) is easy to obtain by measuring the photosynthetically active radiation (PAR) close to the plant with a suitable quantum sensor, I_a is a more complicated measurement, requiring sophisticated equipment such as an 'integrating sphere'. It is, however, possible to estimate I_a of, e.g. an alga with a flat thallus by measuring the light reaching a PAR quantum sensor with and without the thallus covering it (see the next section). If we thus know this **absorption factor** (**AF**), then $I_a = I_i \bullet AF \bullet PAR$, and for photosystem II: $I_a = I_i \bullet AF \bullet 0.5$. In summary, the most common way of presenting the calculation of ETR is thus:

$$ETR = Y \bullet I_i \bullet AF \bullet 0.5 \qquad (8.1)$$

Since I_i is most commonly measured as PAR using the units μmol photons m^{-2} s^{-1}, and since Y is a unit-less fraction of 1.0, as is AF, it follows that the unit derived for ETR is μmol electrons m^{-2} s^{-1}. This measure of photosynthetic rate is both in theory and, luckily, in practice, often stoichiometrically comparable with true rates measured as O_2 evolution (or, again making assumptions, CO_2 evolution), at least under low-light conditions (see the next section). If AF is not known, then a relative value of ETR (rETR) can be derived by simply multiplying Y with I_i. The case for microalgae is a little more complicated and is described below.

8.4 How to measure PAM fluorescence

8.4.1 Macrophytes

The most popular PAM fluorometer for photosynthetic measurements of marine macrophytes (both algae and seagrasses), benthic mats and photosymbiont-containing invertebrates (corals, sponges, etc.) is the Diving-PAM manufactured by Walz, Germany. This fluorometer is enclosed in a water-tight tube and can easily be taken underwater whilst wading, snorkelling or SCUBA diving. The measurement of Y (according to the principle described above) is carried out within 1 s, is displayed, and is stored in the instrument's memory. While the absolute fluorescence unit F (including F_o, F_m and F_m') are of no relevance for the determination of Y and ETR, the manufacturers of PAM fluorometers recommend certain lower and upper limits so as to generate valid values. In addition, we (and co-workers) have found that Y values of <0.1 should be discarded since they are not reliable; this is because F and F_m' are then too close to one another for an accurate separation of the two.

In order to arrive at true ETRs (μmol electrons m^{-2} s^{-1}), the fraction of light absorbed by the leaf or thallus (i.e. the absorption factor, AF) must be known, as must the distribution of photons between photosystems II and I. The latter is often assumed to be 0.5, which is fine for most purposes (but see later in Section 9.5 for some deviations from this value). Regarding AF, we (and co-workers) have found that the simple method of estimating absorbed light described above leads to valid ETRs that are, on a molar (mol:mol) basis, comparable to rates measured as, e.g. O_2 evolution (i.e. ETR equals $0.25 \bullet O_2$; 4 electrons transported through photosystem II per 1 molecule of O_2 evolved, see the next section for the empirical validation of this).

Photosynthesis *vs.* irradiance (*P–I*) curves can be generated by many commercially available PAM fluorometers provided that they can illuminate the plant. The halogen lamp of the Diving-PAM that generates the saturating light for F_m and F_m' determinations, for example, can also irradiate the plant with 8 different actinic light intensities for short (up to a few minutes each) time periods. Since it has been debated how such short-time responses to various irradiances relate to more ecologically relevant changes in irradiance (e.g. diurnally), these *P–I* curves have been termed **rapid light curves**

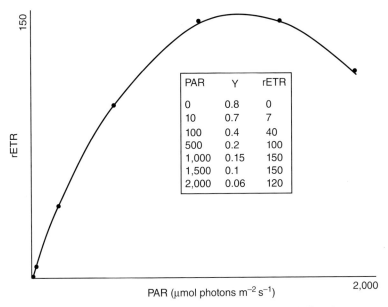

PAR	Y	rETR
0	0.8	0
10	0.7	7
100	0.4	40
500	0.2	100
1,000	0.15	150
1,500	0.1	150
2,000	0.06	120

PAR (μmol photons m^{-2} s^{-1})

Figure 8.8   Relations between yield (Y), **irradiance incident on the thallus** (I_i, measured as photosynthetically active radiation, PAR) **and relative electron transport rates** (rETR). See text for details. Drawing by Sven Beer.

(**RLC**), and their relevance will be touched upon later. The data underlying such curves show a reduction in Y as irradiance increases, but still show an increase in ETR (up to light saturation). In order to understand this, we have prepared the following example, which seemed to be a good basis for understanding the relationship between Y and ETR in general when we teach it to students (you may follow the discussion on the invented scenario of Figure 8.8): An RLC was generated by first measuring Y in virtual darkness (with only the weak measuring light), and the quantum yield for electron transport through photosystem II (Y) was 0.8, i.e. 80% of the photon energy absorbed was used for photosynthetic electron transport. However, since the irradiance from the measuring light beam was virtually nil, the relative rate of electron transport (rETR, = $Y \cdot I_i$, the latter measured as PAR) was close to zero. As irradiance increased, Y decreased but rETR increased up to an irra-

diance of 1000 μmol photons m^{-2} s^{-1}. Since rETR equals $Y \cdot I_i$ or, expressed in another way, $Y \cdot$ PAR, this increase was because PAR increased relatively more than Y decreased within this irradiance range. At 1500 compared to 1000 μmol photons m^{-2} s^{-1}, Y decreased as much as PAR increased (percentage wise), and so rETR remained the same, i.e. photosynthetic light saturation had been reached. At the high irradiance of 2000 μmol photons m^{-2} s^{-1} (~full sunlight), Y decreased drastically, and more so than the increase in irradiance, and so rETR dropped, the phenomenon which we know as photoinhibition.

In order to transform rETR to true rates of ETR (in μmol electrons m^{-2} s^{-1}) the AF has to be known, and the 0.5 distribution factor of absorbed photons between the two photosystems must be 0.5 or close to it, which can usually be assumed to be true. In another example of RLCs, two such curves are depicted in Figure 8.9: one of an imaginary seagrass plant

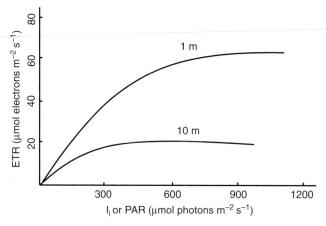

Figure 8.9   Examples of rapid light curves (RLC): Electron-transport rates (ETR) as a function of incident irradiance (I_i or PAR) measured during midday for leaves of the seagrass *Halophila stipulacea* growing at 1 and 10 m depths. See text for details. Adapted after: Beer S, Björk M, Gademann R, Ralph P, 2001. Measurements of photosynthetic rates in seagrasses. In: F.T. Short and R. Coles (eds.) Global Seagrass Research Methods. Elsevier Publishing, The Netherlands. Copyright (2013), with permission from Elsevier.

leaf that grows at 1 m (experiencing 1500 μmol photons m^{-2} s^{-1} during midday) and the other growing at 10 m depth (at 300 μmol photons m^{-2} s^{-1}). As can be seen, the maximal ETR (ETR$_{max}$ or P_{max}) was about 3 times higher for the shallow- than the deep-growing plant, and its light-saturation point (I_{sat}) about twice as high. The AF can be measured by covering the quantum sensor of the Diving-PAM with the seagrass leaf (this should be done under water) and calculating the fraction of PAR absorbed by the leaf. In thin-leaved (or thin-thallused) plants, chlorophyll is the main absorbant of PAR (>90%) as can be determined by measuring the absorption by the whole thallus or leaf and then extracting out all photosynthetic pigments in, e.g. dimethyl formamide and re-measuring the absorption of the now bleached remaining material. In our example, the AF values for the 1- and 10-m growing plants were 0.4 and 0.6 at the time of the measurements, respectively; deeper-growing leaves usually have a higher density of chlorophyll per surface area. (For the specific seagrass measured here, the AF can change also diurnally,

see Section 9.5.) Assuming a 50% division of absorbed photons between the two photosystems, the ETRs for the two plants at, e.g. an irradiance of 300 μmol photons m^{-2} s^{-1} showed Y values of 0.64 and 0.20 and ETRs, calculated as $Y \cdot PAR \cdot AF \cdot 0.5$, of 38.4 and 18.0 μmol electrons m^{-2} s^{-1}, respectively. The same calculations for the other irradiances were the basis for generating the two RLCs depicted in Figure 8.9.

In a final example of RLCs, we wish to show to what degree ETRs from such curves can be used for quantitative measures of photosynthetic rates by comparing them with the more classically measured rates of true O_2 evolution. The experiments leading to these comparisons were performed using an O_2 electrode chambers to which a PAM fluorometer was also connected such that O_2 exchange and ETR could be measured simultaneously (see insert in Figure 8.10a). Thus, after each period of increasing irradiance, after a steady-state O_2 evolution rate had been established, the PAM fluorometer was induced to measure Y, from which ETR was calculated, and this sequence was repeated for different irradiances. True rates of O_2

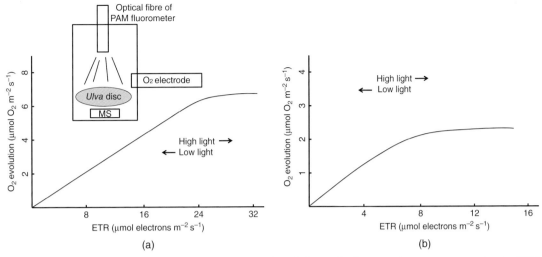

Figure 8.10 **Experimental setup** (insert) and **results of O₂ evolution *vs*. electron transport rate** (ETR) measurements on the green macroalga *Ulva lactuca* (a) and the seagrass *Zostera marina* (b). The setup includes a small, temperature-controlled, chamber in which a circular disk of *Ulva* (depicted) or a seagrass leaf was swirled around by a magnetic stir-bar (MS). See text for details. Reprinted from: Beer S, Larsson C, Poryan O, Axelsson L, 2000, Photosynthetic rates of *Ulva* (Chlorophyta) measured by pulse amplitude modulated (PAM) fluorometry. European Journal of Phycology 35: 69–74. Copyright (2000) with permission from Taylor & Francis (except for the insert) (a), and Reprinted from: Beer S, Vilenkin B, Weil A, Veste M, Susel L, Eshel A, 1998. Measuring photosynthetic rates in seagrasses by pulse amplitude modulated (PAM) fluorometry. Marine Ecology Progress Series 174: 293–300. Copyright (2013), with permission from Inter-Research. (b).

evolution were obtained by correcting apparent rates with the dark respiration obtained directly after each irradiance (as discussed also in Box 8.1). As can be seen, in some cases (e.g. for discs of *Ulva lactuca*, Figure 8.10a), there was an initial linear correlation between O_2 evolution and ETR; this linearity was maintained at irradiances up to 500 μmol photons m^{-2} s^{-1}, where after linearity ceased. Importantly, the O_2/ETR ratio was close to 0.25 within that linear range, i.e. using an AF measured as described above and using the 0.5-factor yielded ETRs that were stoichiometrically related to photosynthetic O_2 evolution. While we find that a molar O_2/ETR ratio of 0.25 at low irradiances can generally be maintenance if a valid way to determine AF values can be devised (similar to what we suggest here for flat, thin-leaved or -thallused plants), the devi-

ation of O_2/ETR from linearity is commonly observed at high irradiances. One such extreme example is given in Figure 8.10b: While thin-leaved seagrasses such as *Halophila* species show linear O_2/ETR ratios up to irradiances of several hundred μmol photons m^{-2} s^{-1}, thicker-leaved ones deviate from linearity at much lower irradiances (as do macroalgae that have a thicker and more complex anatomy than *Ulva*). In the example given, linearity is compromised at *ca.* 100 μmol photons m^{-2} s^{-1}, above which O_2 evolution lags behind ETR. The drop in O_2/ETR ratios at high irradiances has been ascribed to photorespiration or other processes where O_2 is consumed but where ETR is largely maintained. For example, while O_2 is reduced in the water–water cycle, electrons still flow through photosystem II (and part of them are used just for reducing O_2).

8.4.2 Microalgae

Similar approaches to those described above for macroalgae can be used for microalgae. It is possible with dense cultures or biofilms to simply place a fibre optic probe directly into cultures or up against a biofilm or culture vessel and use the same techniques as described above. In some instances, cells can be filtered and the fibre optic placed up against the algal 'artificial leaf' thus formed. It is, however, more usual to use a cuvette system for microalgal suspensions. In some systems, the measuring-light excitation is supplied from a number of different wavelength light-emitting diodes (LED). The Walz (Germany) PhytoPAM Phytoplankton Analyser for instance uses 4 different colours: blue (470 nm), green (520 nm), short-wavelength red (645 nm) and long-wavelength red (665 nm). The different coloured measuring-light pulses are applied sequentially and at a high frequency, so that the instrument measures the chlorophyll fluorescence from all 4 wavelengths almost simultaneously. This can be used in some cases, and with proper calibration against known species, to determine approximate contributions of different algal classes with different types of light-harvesting pigments in a mixed population. Thus, for diatoms and dinoflagellates excitation is greater by the blue (470 nm) and green (520 nm) beams due to the presence of fucoxanthin/peridinin, whereas cyanobacteria show strongest excitation at 520 and 645 nm, reflecting the importance of the phycobiliproteins in their light harvesting. A significant issue with PAM-based measurements on suspensions is estimating the AF. Ideally, this is measured using an integrating sphere but absorptance can be measured by collecting cells from suspension onto a glass fibre filter, placing this filter close to the detector of a spectrophotometer, with an opal glass between sample and detector to provide diffuse light, and running

an absorbance spectrum. The filter is then extracted with methanol or another suitable solvent and the scan repeated. The difference in integrated light between the two treatments is the absorbed light.

An alternative approach to PAM fluorescence measurements is to use Fast Repetition Rate Fluorometry (**FRRF**). Here, instead of a modulated measuring beam and a relatively long (typically 800 ms), 'multiple-turnover' saturating flash, the sample is exposed to a series (perhaps up to 100) of very short ($\sim$1 µs), intense pulses (flashlets) at a high frequency (0.35 MHz). These progressively populate reaction centres of photostem II and reduce Q_A and as a consequence fluorescence values are gradually raised to F_m (Figure 8.11). After the first $\sim$280 µs, samples are then exposed to a set of 20 or so flashlets at lower frequency (typically around 20 Hz) that allows determination of the re-oxidation kinetics for the Q_A pool. Applying algorithms to the data allows calculation of the usual parameters such as F_m and F_o but also the effective cross-sectional area, which in turn then allows for estimate of actual rates of electron transport (see Huot and Babin, 2011).

Yet another approach to estimate the primary productivity of the oceans is by remote, **satellite**, **monitoring** (see the below box by Stewart Larsen).

8.5   What method to use: Strengths and limitations

When we tell students the pros and cons of various marine-photosynthetic measurement techniques, they often ask us which one to choose. Therefore, this summary of strengths and limitations is intended to answer some of those questions. Largely, the method of choice depends on what one wants to measure: If a growth-related measurement is sought, then

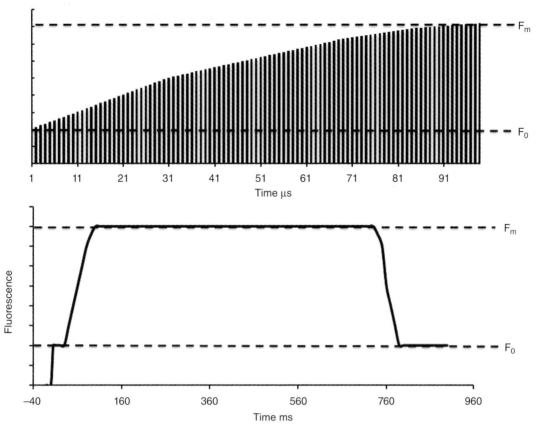

Figure 8.11   A schematic **comparison of FRR** (upper part, FRRF) **and PAM** (lower part) **fluorometry**. In FRRF, electron carriers are gradually closed as a result of many single turnover flashlets, each of ~1 μs duration, leading from F_0 to F_m. In PAM fluorometers, the saturation pulse is much longer (note the different X-axis scales in the 2 plots) and as a result is a consequence of multiple turnovers over the 800 ms duration of the flash. Drawing by John Beardall.

gas exchange is a natural choice, and O_2 is easier to measure in water than CO_2. If photosynthetic responses to various environmental conditions (especially pertaining to irradiance) are sought, then PAM fluorometry is a quick and exact way both in the laboratory and *in situ*.

The big minus of gas-exchange measurements is the general need for enclosures (see, however, 'open system' gas exchange in Section 12.3). In the laboratory, such enclosures are generally of small volumes (i.e. a few ml) and when measuring macrophyte photosynthesis then these plants often have to be cut

up to fit into the enclosures. This itself is a problem (see the section 'Using whole plants' below), but another problem is that both the enclosure (be it glass or some transparent plastic) and the plants therein change the environment from that of natural, open, systems. Thus, increasing O_2 levels within enclosures where photosynthesis exceeds respiration may favour photorespiration (see Section 7.4), and decreasing Ci levels may restrict photosynthesis (Section 11.1). Further, increases in pH may shift the Ci equilibria away from CO_2 and, thus, restrict photosynthesis further, especially

Box 8.2   Satellite measurements of chlorophyll

By **Stuart Larsen**, Monash University, AU (stuart.larsen@monash.edu)

In contrast to traditional methods of measuring chlorophyll, satellites provide many unique advantages. First, these include a global coverage with access to regions that are otherwise unreachable. Secondly, large areas, typically 1000–2000 km wide, are able to be sampled at once, providing a snapshot where all the values are at the same point in time and, with orbits of around 100 min, the whole globe can be sampled every day. In contrast, ship-based observations of a similar transect may be separated in time by several days, even if multiple ships are used. Thirdly, a single instrument system makes all the measurements, therefore requiring only one set of calibrations. Changes in instrument quality over time can be monitored and corrected, and improved algorithms can be applied to prior data. Finally, in terms of any other way of obtaining the same sort of data coverage, satellites are very economical.

To date, a number of satellite platforms have carried a suite of different instruments including the Coastal Zone Colour Scanner (CZCS) 1978–1986, the Sea-viewing Wide Field-of-view Sensor (SeaWiFS) Aug 1997–2010, and the Moderate Resolution Imaging Spectroradiometer (MODIS) 1999-present, yielding a vast amount of data. The interpretation of these observations is more complicated, however.

The very high absorbance of long wavelengths by water prohibits the use of fluorescence-based measurements of chlorophyll, as a proxy for photosynthesis, over the oceans. Instead, chlorophyll concentration is determined from the ratio of reflected green light (555 nm) to one or more wavelengths of blue light (443, 490 and 510 nm) depending on the satellite. Over the open ocean, with low chlorophyll concentrations (typically 0.2 mg m^{-3}), most of the reflected light is in the blue. As chlorophyll concentration increases, more blue light is absorbed while more green light is reflected from the oceans, and in high-chlorophyll regions the green reflectance dominates over the blue. An empirically based algorithm, derived from surface optical and chlorophyll measurements over the world's oceans, is then used to convert this ratio into a chlorophyll concentration. The algorithm factors are corrected for latitude and uses assumptions for species type and cell size for the different oceanic regions. These algorithms work best over the oligotrophic oceans, but decreasingly well as chlorophyll concentrations increase. In part this is because in highly productive coastal seas, with high chlorophyll concentrations, the input of suspended sediment, organic matter and stratification from freshwater inputs all combine to increase the uncertainty. Light reflected from some particle within the ocean must travel a distance at least twice the depth of that particle, and this limits the determination of chlorophyll concentrations by satellite to just the near-surface waters.

While a computer algorithm will always yield a result, given some input data, the assumptions on which that data are based may not remain constant in a changing climate. Stratification may alter the mean depth of reflectance and so spectral ratios of the water leaving irradiance, as well as the light and nutrient environment in which the phytoplankton grows, and so the chlorophyll content of the cells. Changes in species (which have different pigment compositions) or cell size may also lead to apparent changes in chlorophyll that are not necessarily reflected in biomass. Thus, while the need for satellite measurements of the global ocean is greater than ever in order to better understand the effects of global climate change, caution is required to not blindly accept the data they provide without continued surface checking that the algorithms used remain valid in a changing ocean environment (Dierssen, 2010).

Satellite data now permits global-scale studies to be undertaken, for example the effect of hemispheric phenomena such as El Niño on chlorophyll patterns (Behrenfeld et al., 2006). At the other end of the scale, as additional wavelength channels continue to be added, and camera resolution improves, the detection and near real-time monitoring of harmful algal blooms near the coast (Klemas, 2012) or coccolithophore blooms in the Southern Ocean (Moore et al., 2012) are now possible.

Some useful references for the interested reader: Behrenfeld MJ, O'Malley RT, Siegel DA, McClain CR, Sarmiento JL, Feldman GC, Milligan AJ, Falkowski PG, Letelier RM, Boss ES, 2006. Climate-driven trends in contemporary ocean productivity. Nature 444: 752–755; Dierssen HM, 2010. Perspectives on empirical approaches for ocean color remote sensing of chlorophyll in a changing climate. Proceedings of the National Academy of Sciences 107: 17073–17078; Klemas V, 2012. Remote sensing of algal blooms: An overview with case studies. Journal of Coastal Research 28: 34–43; Moore TS, Dowell MD, Franz BA, 2012. Detection of coccolithophore blooms in ocean color satellite imagery: A generalized approach for use with multiple sensors. Remote Sensing of Environment 117: 249–263.

in less-efficient HCO_3^- users. While this is true both in the laboratory and for *in situ* incubations, the latter situation may allow for choosing both a lower biomass/volume ratio and closer to natural situations in terms of, e.g. stirring. These influences by incubation chambers were earlier called 'bottle effects', and they are still as troublesome as always.

While PAM fluorometry does not require enclosures, one of its main restrictions is that it measures rates only of true photosynthesis. While this can also be an advantage if such rates are desired, the disadvantage over gas-exchange measurements is that it is rather far-fetched, if not impossible, to arrive at productivity estimates since the component of respiration is not included in the measurement. Therefore, this method is more suitable when photosynthetic responses *per se* are sought, e.g. as a function of various environmental conditions. We find that for macrophytes, PAM fluorometry is especially suitable for elucidating photosynthetic responses and acclimations to

irradiance (see following sections). For phytoplankton, the traditional methods for determination of photosynthesis have been ^{14}C incorporation or oxygen exchange using Winkler titrations, though these can suffer from bottle effects so experiments need to be designed, and data interpreted, carefully. The PAM fluorometric methods are generally too insensitive for use with many natural populations, especially in oligotrophic waters, and usually only allow determination of rETR rather than true rates of electron transport. However, FRRF is ideal for use with phytoplankton and does provide real electron-transport rates.

8.5.1 Rapid light curves

It has been debated how rapid light curves (RLC) relate to natural changes in irradiance. One argument against them is that the plants do not have time to acclimate, i.e. develop their potential responses, to such rapid changes in

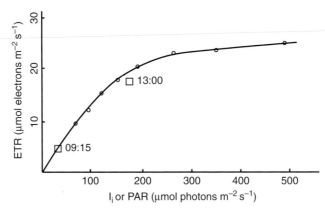

Figure 8.12   Rapid light curve (RLC, circles**) and average electron transport rates** (ETR) of 20–30 point measurements (squares) measured for the seagrass *Halophila stipulacea* growing at 6 m depth. The RLC was measured at noon-time and the point measurements were performed at 09:15 (lower left square) and 13:00 (upper middle square). Adapted after: Beer S, Vilenkin B, Weil A, Veste M, Susel L, Eshel A, 1998. Measuring photosynthetic rates in seagrasses by pulse amplitude modulated (PAM) fluorometry. Marine Ecology Progress Series 174: 293–300. Copyright (2013), with permission from Inter-Research.

irradiance as generated by the RLCs (minutes rather than the hourly changes over a day). In order to partly overcome such limitations, we recommend a few things that should be kept in mind when using them: First, the RLCs should be started immediately after inserting a thallus or leaf into or onto the device holding it whilst the RLC is being performed, i.e. if the RLC starts with darkness, followed by increasing irradiances, then the plant taken from a certain irradiance environment should not be given time to dark-adapt. Secondly, the approximate midpoint irradiance of the RLC should correspond to the ambient irradiance from which the plant was taken. If those rules are followed, then a good relationship can be obtained between the photosynthetic performance predicted from the RLC and the actual response during ambient conditions obtained by what we call point measurements. (A point measurement is a single measurement of Y (and, simultaneously, PAR) under a certain environmental condition that yields the true ETR under that condition.) An example of this is given in Figure 8.12: A RLC was performed

at noontime on a leaf of a seagrass growing at 6 m depth (the circles indicate the measurements of ETR at the 8 irradiances). During the same day, 20–30 point measurements of ETR were performed at 09:15 and 13:00 (left and right squares, respectively) at the average ambient irradiances indicated. As can be seen, the results of the latter closely relate to the results of the RLC at similar irradiances. What this exercise shows is that a RLC, when performed as described above, can reflect the response to irradiance at a different time scale: Here, the RLC performed during midday for a duration of <2 min reflects the plant's response to irradiance throughout the day.

8.5.2 F_v/F_m

As mentioned above, the maximal yield (Y) value F_v/F_m can be used as a measure of stress that affects components of photosystem II, such as the D1 protein. Since F_v/F_m is the highest value of Y, and since we have seen that Y decreases with irradiance, it follows that F_v/F_m must, by definition, be measured in darkness. However, a short period of darkness may not be enough in order to arrive at a representatively high and stable value of F_v/F_m; a very short (< 1 s) period of darkness will result in what we call Y_0 (see the next section). If several hours of darkness precedes the measurement (i.e. a whole night), then F_v/F_m can be a good measure of more chronic stress, including water or desiccation stress (in the intertidal), thermal stress and, also, the light stress prevailing during prolonged conditions of, e.g. high irradiances. For example, the higher chronic photoinhibition found in shallow-growing than in deep-growing corals (see Section 9.3) is manifested in a lower F_v/F_m value in the former as measured in the early morning, pre-dawn. If shorter-term dark-adaptations are used (e.g. the 10–15 min normally recommended), then shorter-term stresses are measured before the sensitive components of PSII have been relaxed or repaired, e.g. the daily down-regulation of photosynthesis by high midday irradiances. Thus, the type of stress measured as F_v/F_m will warrant the time for dark-adaptation before the measurement. Time courses of dark-adaptation followed by F_v/F_m measurements are therefore an excellent means to arrive at the correct time for dark-adaptation depending on the type of stress to be measured.

8.5.3 Alpha, "uses and misuses"

When analysing photosynthesis-irradiance (P–I) curves derived from RLCs such as depicted in Figure 8.12, the initial slope of the curve, often termed alpha (α), is used as a proxy of the maximal quantum yield. The problem with this procedure is that in order to arrive at quantum yields, the photosynthetic output (in this case ETR) should be measured per absorbed photons, but the P–I curves are almost always presented per the incident photon flux of PAR as measured near a leaf or a plant assemblage. Thus, even if used relatively when comparing, e.g. initial slopes of plants from different depths or during different seasons, the quantum yields derived from those slopes will differ if, e.g. the absorption factor (AF) is different. For example, many plant species contain more chlorophyll per leaf surface area when the ambient irradiance is lower, e.g. at depth or during winter months. If this is not taken into account, then the maximal quantum yield will be underestimated (since the absorbed light is always less than the incident light, and so ETR will be calculated per a higher irradiance), and grossly so for plants under the low-light conditions. This can be illustrated as in Figure 8.13: When the initial slope for *Ulva* sp. was measured using ETR *vs.* incident irradiance (I_i, as is usually done), then there was a 3.3 times higher initial slope for the winter *vs.* summer algae. If, however, the initial slope was measured using the absorbed irradiance (I_a), then the slope was only 1.2 times higher. This is simply because the winter plants had a much higher chlorophyll content, and thus AF, than the summer ones, and of course the rationale for using I_a when relating initial slopes to quantum yields is that the latter relates to absorbed photons. If only incident light is used, rather than absorbed, then α is simply a reflection of the efficiency of light harvesting. Further, since ETR does take into account absorbed light (i.e. is calculated including the AF value, see Equation 8.1 above), then of course the x-axes of P–I curves from which quantum yields are to be derived should also display absorbed light. If this was unclear, then perhaps the following explanation will clear things up a little (look at Figure 8.13): The usual way of presenting RLCs is by depicting ETR *vs.* incident irradiance (I_i, panel a), and when the initial slope of such curves are used to determine the maximal quantum yield, then there is an

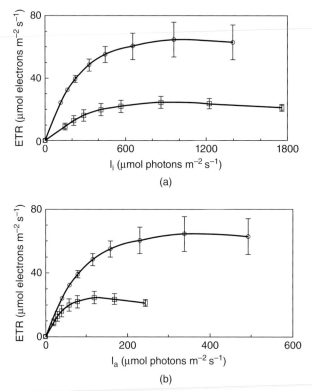

Figure 8.13   Differences between rapid light curves (RLC) **measured in December** ('low-light' *Ulva*, AF = 0.70, circles) **and July** ('high-light' *Ulva*, AF = 0.27, squares) based on incident irradiance (I_i) (a) and absorbed irradiance (I_a) (b). Reprinted from: Saroussi S, Beer S, 2007. Alpha and quantum yield of aquatic plants as derived from PAM fluorometry. Aquatic Botany 86: 89–92. Copyright (2013), with permission from Elsevier.

apparent large difference between, in this case, winter (low-light *Ulva*, upper curve) and summer (high-light *Ulva*, lower curve). If, however, it is realised that ETR should be calculated per absorbed light (and quantum yields also relate to absorbed light), and so ETR is depicted *vs.* absorbed irradiance (I_a, panel b), which is the correct way if quantum yields are sought, then the result shows that the difference in maximal quantum yield between winter and summer plants is much less pronounced. This is because the difference in Panel A is diminished when taking into account that the summer plants contain a much lower concentration of chlorophyll and, so, the I_a on the x-axis acquires a relatively lower value compared to

I_i, resulting in a higher initial slope and maximal quantum yield than if this lower absorption were to be ignored. (If this is still not clear, then we apologise!)

Another anomaly when presenting the increasingly popular RLCs is that maximal quantum yields are calculated from their initial slopes. It seems that those who do so forget that PAM fluorometry measures, directly and almost instantaneously, just quantum yields, without the need for displaying or calculating initial slopes. If so, the maximal quantum yield is the Y value displayed by the PAM fluorometer when the plant is in darkness, just before the RLC starts to illuminate it with the different irradiances. The fact that such Y

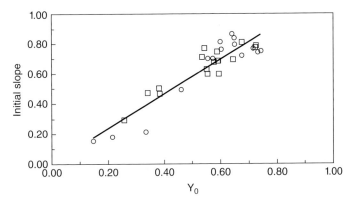

Figure 8.14　Correlation between the initial slope of rapid light curves (RLC) (mol electrons mol photons^{-1}) **using absorbed irradiance** (I_a) as the independent variable, **and _Y_ derived from the initial step of the RLC** (Y_0) generated in December ('low-light' _Ulva_, circles) and July ('high-light' _Ulva_, squares, at different times of the day). Each data point represents one RLC. Reprinted from: Saroussi S, Beer S, 2007. Alpha and quantum yield of aquatic plants as derived from PAM fluorometry. Aquatic Botany 86: 89–92. Copyright (2013), with permission from Elsevier.

values (here called Y_0) really represent maximum quantum yields was shown for, again, _Ulva_ sp. by comparing them with quantum yields obtained from ETR _vs._ irradiance curves based on I_a (Figure 8.14). Based on this and similar results, our recommendation is to determine maximal quantum yields by PAM fluorometry by simply measuring _Y_ in a darkened plant either after prolonged darkness (Y_0 will be high in a dark-adapted plant) or within _ca._ 1 s of darkness following a certain light treatment (the higher the light was before the darkening, the lower Y_0 can be expected to be).

8.5.4　Using whole plants

In addition to the general limitations of 'bottle experiments', whether in the laboratory or in the field, it may also be necessary to keep macrophytic plants intact for valid estimations of photosynthetic rates. A good example can be given for seagrasses: These higher plants have roots, rhizomes and shoots (including leaves and, sometimes, flowers), and it may be important to keep those structures intact dur-

ing measurements of, e.g. photosynthetic rates. Before PAM fluorometry became popular, and given the difficulty of incubating seagrass leaves for _in situ_ gas-exchange measurements while changing, e.g. Ci concentrations around the plants, seagrass leaves were taken to the laboratory for such measurements. There, given the small volumes of the O_2 measurement chambers, the leaves had to be cut down to a fraction of their original size in order to fit into those chambers. Results of such measurements are depicted in Figure 8.15, where the photosynthetic O_2 response to Ci was appraised by adding increasing amounts of HCO_3^- into the chamber. As can be seen, the seagrasses responded favourably to Ci concentrations above that of normal seawater (2100 µM), while the macroalgae also tested did not. From such experiments, it was deduced that while most macroalgae were saturated by today's seawater Ci concentration, seagrasses were not, and so would respond favourably to future global increases in CO_2 concentrations, which would partly be reflected in a higher Ci concentration also of the oceans. From such observations it was even deduced in a

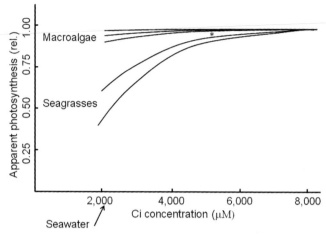

Figure 8.15   Photosynthetic O₂ exchange responses of three macroalgae (the green *Ulva*, the red alga *Palmaria* and the brown alga *Laminaria*, upper 3 lines) **and two seagrasses** (*Thalassia testudinum* and *Zostera marina*, upper and lower curve, respectively) **to increased Ci concentrations**. Reprinted from: Beer S, Koch E, 1996. Photosynthesis of seagrasses vs. marine macroalgae in globally changing CO₂ environments. Marine Ecology Progress Series 141: 199–204. Copyright (2013), with permission from Inter-Research.

paper co-authored by SB that seagrasses would gain an advantage over macroalgae and would outcompete the latter (Beer and Koch, 1996). This thought was challenged by John Raven who, in his peer-review of that paper, claimed that this was not to be since seagrasses grow on soft sediments and macroalgae on hard rocks (with very few exceptions, e.g. some *Caulerpa* species can grow on soft sediments too), and so there was no competition for space between the two plant groups. From that point on, I (SB) have looked hard for places where macroalgae and seagrasses do grow together, and have found quite a few, see, e.g. Figure 8.16!

We are getting carried away now: The topic here is the need to use intact plants for measurements of photosynthesis. Results such as those presented in Figure 8.15 have led to the dogma that seagrasses are Ci limited in today's oceans, and will benefit from additional Ci as global CO₂ concentrations increase. However, when similar Ci-enrichment experiments were carried out *in situ* using PAM fluorometry, it

was found that not all seagrasses were Ci limited. For example, two tropical seagrasses growing in shallow water (i.e. high irradiances) did not respond to increasing Ci concentrations *in situ*. Such results contrast with previous ones

Figure 8.16   The seagrass *Phyllospadix* (in the background) **and a brown kelp** (in the foreground) growing together in the Pacific of northern Mexico. While the kelp is anchored to the underlying rocky substrate, the seagrass roots grow in crevices filled with soft sediments. Photo by Sven Beer.

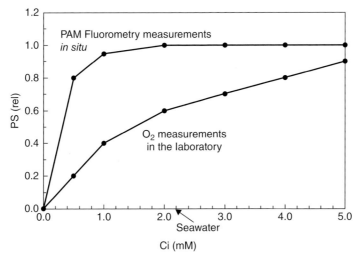

Figure 8.17   Principal differences that can occur **when photosynthetic responses** to inorganic carbon (Ci, here in mM) **are measured in a small O₂ electrode chamber** in the laboratory **and**, for the same species, **using PAM fluorometry in situ** on intact plants from the same natural population as used for the O₂ measurements. Drawing by Sven Beer.

Table 8.1   Possibilities when **using various techniques** for measuring photosynthetic rates in marine plants.

Technique	O₂ exchange	CO₂ exchange	Fluorescence	
			PAM fluorometry	FRRF
Characteristic				
True *vs.* apparent rates	apparent	apparent	true	true
'Dark' respiration	measurable	measurable	not measurable	not measurable
Response time	minutes to hours	minutes to hours	<1 s	<1 s
Units measured	$\mu mol\ O_2\ m^{-2}\ s^{-1}$	$\mu mol\ CO_2\ m^{-2}\ s^{-1}$	$\mu mol\ electrons\ m^{-2}\ s^{-1}$	$\mu mol\ electrons$ $mol\ chl^{-1}\ h^{-1}$
Enclosures needed	yes (usually)	yes (usually)	no	no
In situ application	difficult	possible	easy	easy
Best measured in (air/water)	water	air	water and air	water

where Ci-limitation was found under laboratory conditions; this difference is principally illustrated in Figure 8.17. We do not know if marine plants such as seagrasses suffer from their leaves being cut and trimmed into small O₂ chambers, and then measured under laboratory conditions, but drawing interferences from terrestrial higher plants it is likely that they do. In all, we recommend that whenever possible photosynthetic measurements be carried out *in situ* under as close to natural conditions as possible.

In summary of this section, some possibilities when using the various techniques described above, including their pros and cons, are listed in Table 8.1.

The brown alga *Halidrys siliquosa* with epiphytes at the Swedish west coast. Photo by Katrin Österlund.

Chapter 9

Photosynthetic responses, acclimations and adaptations to light

9.1 Responses of high- and low-light plants to irradiance

When performing short-term photosynthetic gas-exchange measurements as a function of irradiance (*P–I* curves, see Chapter 8 above) on marine plant species that grow in **high-** *vs.* **low-light environments**, then the typical difference between their **apparent** photosynthetic responses are as illustrated in Figure 9.1a. The 'typical' explanations behind these differences are as follows: **Alpha** (α), which is a measure of the maximal photosynthetic efficiency (or quantum yield, i.e. photosynthetic output per photons received, or absorbed (depending on how irradiance is measured and expressed, see the section Alpha, "uses and misuses" in Section 8.5) by a specific leaf/thallus area, is high in low-light plants because pigment levels (or pigment densities per surface area) are high. In other words, under low-irradiance conditions where few photons are available, the probability that they will all be absorbed is higher in plants with a high density of photosynthetic pigments (or larger 'antennae', see Chapter 5 and Section 9.1 and below). In yet other words, efficient photon absorption is particularly important at low irradiances, where the higher concentration of pigments potentially optimises photosynthesis in low-light plants. In high-irradiance environments, where photons are plentiful, their efficient absorption becomes less important, and instead it is reactions downstream of the light reactions that become important in the performance of optimal rates of photosynthesis. The CO_2-fixing capability of the enzyme Rubisco, which we have indicated as a bottleneck for the entire photosynthetic apparatus at high irradiances, is indeed generally higher in high-light than in

Photosynthesis in the Marine Environment, First Edition. Sven Beer, Mats Björk and John Beardall.
© 2014 John Wiley & Sons, Ltd. Published 2014 by John Wiley & Sons, Ltd.
Companion Website: www.wiley.com/go/beer/photosynthesis

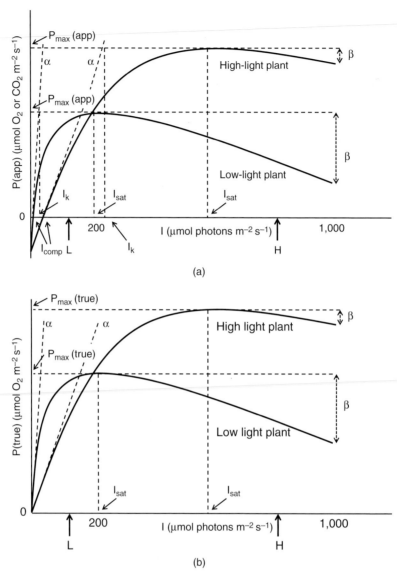

Figure 9.1   Apparent (a) **and true** (b) **photosynthetic responses to irradiance typical for high- and low-light growing marine plants** (see text regarding Figure 8.1 for abbreviations of terms). Among the differences we note that low-light plants as compared with high-light ones have lower maximal photosynthetic rates at light saturation (P_{max}) and lower light saturation irradiances (I_{sat}), but higher maximal photosynthetic efficiencies (α) and higher extents of photoinhibition (β). However, such photoinhibition is seldom encountered by low-light plants in nature since they typically live at low irradiances (*L*); high-light plants typically live at the higher irradiances (*H*). Note that the only difference between panels a and b is that the photosynthetic rates in the latter have been corrected for the rate of dark respiration (which in this example was equal for both plant forms though is frequently higher for high-light grown plants). Drawing by Sven Beer.

low-light plants because of its higher concentration in the former. So, at high irradiances where the photon flux is not limiting to photosynthetic rates, the activity of Rubisco within the CO_2-fixation and -reduction part of photosynthesis becomes limiting, but is optimised in high-light plants by up-regulation of its formation. This is also why P_{max} is higher in the high-light plants, as is I_{sat}. These statements about high- and low-light plants, as well as the ones in the next paragraph, are rather dogmatic, i.e. the rationales for the responses measured are often conjecture, and further research should elucidate the exact mechanisms beyond them. However, as an apparent result of the above, α is generally higher in the low-light plants when photosynthesis is expressed on a surface area or dry weight (or cellular) basis. If photosynthesis is expressed per unit chlorophyll, then α values in high- and low-light grown plants are similar (because of their lower and higher chlorophyll contents, respectively).

The above photosynthetic responses have often been explained in terms of adaptation to low light being brought about by alterations in either the number of 'photosynthetic units' or their size (see Figure 9.2). The photosynthetic unit is here, again, thought of as photosystem I (PSI) and photosystem II (PSII), including the antennae pigments serving the reaction centres and the apparatus for electron transport. In the case of shade adaptation involving increases in the number of photosynthetic units in a cell, both antenna size and reaction centre numbers per cell increase in concert, whereas when the photosynthetic unit size increases, only the antenna size, serving the same number of reaction centres, increases (though at very low light decreases in Rubisco levels to reduce energy expenditure on protein synthesis, and thus P_{max}, may also occur). This has consequences for the shape of the P–I curves (shown in Figure 9.3). There are good examples of both strategies occurring in dif-

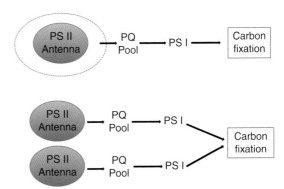

Figure 9.2   Adaptation to low light through, upper panel, increasing the **size of the photosynthetic unit** by increasing antenna pigments (shown for PSII by the dashed ellipse) or, lower panel, by increasing the **number of photosynthetic units** per cell. These changes have consequences for the shape of the P–I curves (Figure 9.3). Drawing by John Beardall.

ferent species of algae (macro- and micro-), as detailed and summarised by Richardson and co-workers (1983).

Finally, the generally larger β in low-light plants reflects their higher sensitivity to high irradiances; i.e. they become more photoinhibited.

In order to express the rates of apparent photosynthesis as a function of irradiance as '**true**' photosynthesis, the rate of dark respiration needs to be corrected for, and this results in the P–I curves of Figure 9.1b. Both α and β attain the same values when calculated on the apparent or true photosynthetic rate curves, as do I_{sat} and I_k (the latter not shown in panel b). (See the beginning of Chapter 8 for the nomenclature of P–I curves.) Note also that the compensation point for light (where the photosynthetic gas-exchange rate equals that of respiration, I_{comp}) can only be given for apparent rates of photosynthesis (i.e. in panel a).

The light responses depicted in Figure 9.1 are those of single thalli or leaf surfaces. However, we may be interested in light responses of an entire assemblage of plants such as a

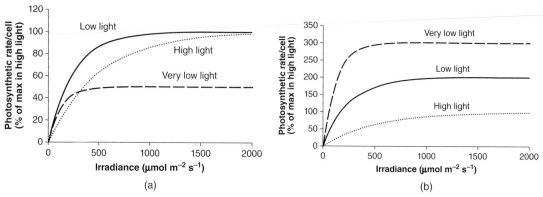

Figure 9.3 Model of the **effects of changing photosynthetic unit size** (a) or **photosynthetic unit number** (b) **on true photosynthetic rate** per cell. Increasing photosynthetic unit size increases photosynthetic rates at sub-saturating irradiances (and at very low irradiances a decrease in Rubisco levels is seen as decreased P_{max}). With increasing photosynthetic number, all aspects of the photosynthetic functional apparatus are enhanced as light for growth is decreased. Drawing by John Beardall.

seagrass meadow or an algal belt (see, e.g. Section 12.3). If so, then it is obvious that less light penetrates to the lower than the outer part of the assemblage such that the 'average' plan will receive less light than measured by a sensor above the canopy. The differences in irradiance between single leaves or thalli *vs.* the sum of all 'green' (i.e. photosynthesising) parts of the plants within the canopy can be illustrated as in Figure 9.4. Thus, for example when the top layer of such a plant assemblage is light saturated (already at less than half of full sunlight, as represented by the single leaf response), the assemblage as a whole is far from light saturated at such an irradiance. In extreme cases, it may be that the top layers on the one hand become photoinhibited or even photo-damaged, but on the other hand they protect the underlying plant parts from such negative effects by shading them.

Now a few words about **photoinhibition**. In general, photoinhibition can be defined as the lowering of photosynthetic rates at high irradiances. This is mainly due to the rapid (sometimes within minutes) degradation of a protein important for the function of PSII, the **D1**

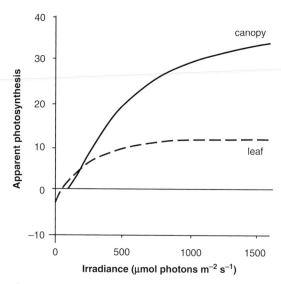

Figure 9.4 The response of single leaves and a canopy containing those leaves to irradiance (see the above text for explanations). This example is taken from spruce trees, but applies also to other plants such as marine macrophytes. Adapted from: Jarvis PG, Leverenz JW, 1983. Productivity of temperate, deciduous and evergreen forests. In: Encyclopaedia of Plant Physiology, new series Vol. 12D. (Ed. by O.L. Lange, P.S. Nobel, C.B. Osmond and H. Ziegler). Springer-Verlag, Berlin. Copyright (2013), with permission from Springer.

protein. If its degradation is faster than its re-formation, then this will cause photosynthesis to slow down, resulting in the photoinhibition termed β in, e.g. Figure 9.1. Before such destructive measures occur, however, there are defense mechanisms that **divert excess light energy** to processes different from photosynthesis; these processes thus cause a **down-regulation** of the entire photosynthetic process while protecting the photosynthetic machinery from excess photons that could cause damage. One such process is the xanthophyll cycle.

The **xanthophyll cycle** has been well described for terrestrial, higher, plants (e.g. by Blankenship). In seagrasses, green algae and brown algae, it is the xanthophyll violaxanthin that is converted, in two de-epoxidation steps, to antheraxanthin, and then to zeaxanthin by the enzyme violaxanthin de-epoxidase which is bound to the lumen side of the thylakoid membrane; zeaxanthin has the property of dissipating excess light energy as heat. This de-epoxidation of violaxanthin to zeaxanthin is triggered by low pH values in the lumen, which result from intensive proton transloca-tion across the thylakoid membrane at high light. Under low light and higher pH (~7.5), the reverse reaction sees zeaxanthin converted back to violaxanthin in a reaction catalysed by zeaxanthin epoxidase. Such a diversion of energy, together with other forms of down-regulation of photosynthesis, can be traced by PAM fluorometry, where it is expressed as non-photochemical quenching, or NPQ, of the fluorescence signal (see Section 8.3).

Another way of diverting excess light energy, especially under conditions where CO_2-fixation and -reduction is limited by a lack of CO_2 (e.g. at high seawater pH values, see Chapter 11), is for the photosynthetically generated electrons from water to reduce O_2 instead of $NADP^+$ and, ultimately, CO_2. Thus, e.g. the Mehler reaction, as part of the water-water cycle, may divert those electrons towards the formation of H_2O_2 and, ultimately, water (see Section 5.4). This way of dissipating excess energy will down-regulate photosynthesis further, but it cannot be detected by PAM fluorometry since it involves the transport of electrons through PSII.

It has also been suggested that the activity of the CCM in marine plants (see Chapter 7) can be a source of energy dissipation. If CO_2 levels are raised inside the cells to improve Rubisco activity, some of that CO_2 can potentially leak out of the cells, and so raising the net energy cost of CO_2 accumulation and, thus, using up large amounts of energy (see also Box 11.1). Indirect evidence for this comes from experiments in which CCM activity is down-regulated by elevated CO_2; under such conditions, and when exposed to high light, levels of NPQ increase, implying that the cells need to up-regulate processes such as the xanthophyll cycle to compensate for the decreased contribution of the CCM to energy dissipation. This phenomenon has been reported in both microalgae and macroalgae.

It is often hard to understand the interaction, and/or distinction, between **photoprotection** and photoinhibition and how these processes affect the photosynthetic response to irradiance. In Figure 9.1, one may, e.g. wonder how the maximal photosynthetic rate (P_{max}) is higher in the high-light plant when down-regulation is in effect? The answer is (probably) that without down-regulation, photoinhibition would be excessive, and that down-regulation allows for higher photosynthetic rates than in the low-light plants, the latter of which is severely (and chronically, see below) photoinhibited at high irradiances in the lack of down-regulation. While high- and low-light plants behave in principle as depicted in Figure 9.1, photosynthesis is in the high-light plants probably down-regulated by the xanthophyll cycle and electron diversion such that photoinhibition is minor. On the other hand, if the low-light plants lack these systems of energy dissipation, then photoinhibition will be greater.

Photoinhibition is often divided into **dynamic** and **chronic** types, i.e. the former is quickly remedied (e.g. during the day, see the example of shallow-growing corals in Section 9.3) while the latter is more persistent (e.g. over seasons, also see Section 9.3). The photoinhibition term β in Figure 9.1 is expressed as a short-term (minutes) response when plants are experimentally exposed to increasing irradiances. In nature, such rapid fluctuations in irradiance may occur, e.g. in intertidal environments where submergence caused by waves alternates with emergence. More commonly, changes in irradiance occur successively during the day, from sunrise, through midday and then towards the sunset. In intertidal (as well as shallow subtidal) benthic plants, these oscillations may be augmented by tidal fluctuations; in some areas the tidal ranges can be >10 m! Therefore, the mechanisms for down-regulating photosynthesis by diverting photon energies and the reducing power of electrons away from the photosynthetic systems, including the possibility of detoxifying oxygen radicals, is important in high-light plants (that experience high irradiances during midday) as well as in those plants that do see significant fluctuations in irradiance throughout the day (e.g. intertidal benthic plants). While low-light plants may lack those systems of down-regulation, one must remember that they do not live in environments of high irradiances, and so seldom or never experience high irradiances. Looking again at Figure 9.1, and remembering that the photosynthetic responses to irradiance depicted are those measured artificially during rather short-term (minutes) of exposure to the different irradiances, we now envisage that low-light plants grow naturally at low irradiances, e.g. at that marked by L during

midday and below it before and after noon, and that high-light plants may grow at a midday irradiance as marked by H. Realising this, we then realise that low-light plants have adapted to low irradiances, and that they may actually have a higher rate of photosynthesis in such low-light conditions than high-light plants. On the other hand, high-light plants may thrive under high-light conditions where low-light plants would be severely photoinhibited.

Two additional features of Figure 9.1 can be mentioned here. First, it is not incidental that the highest irradiance given in the graphs is 1000 μmol photons m^{-2} s^{-1}, which is about half of the maximal irradiance during a sunny summer day. This indicates that even high-light marine plants cannot fully utilise the high irradiances in, e.g. the intertidal or very shallow waters. Thus, often even high-light growing marine plants are considered 'low-light' plants because of their response to (or non-response to, or even inhibition by) high irradiances. Secondly, as already said above, but being so important that it will not do harm to say it again, it must be remembered that the light responses commonly depicted in graphs such as Figure 9.1 are those obtained when taking a high- or low-light plant and subjecting it to artificially generated, often non-natural, irradiances. If, however, we view the high-light plant to grow at a midday irradiance of 800 μmol photons m^{-2} s^{-1} and the low-light one at 100 μmol photons m^{-2} s^{-1}, then it becomes apparent that while the high-light plant indeed performs better at 800 μmol photons m^{-2} s^{-1}, the low-light plant performs better than the high-light plant at the lower irradiance. Thus, there is really an advantage of adapting to low-light conditions for plants that naturally grow in low irradiances. If plants had a mind, one could say that it was worth it for them to invest in pigments, but unnecessary to invest in high amounts of Rubisco, when growing under low-light conditions, and necessary for high-light growing plants to invest in Rubisco, but not in pigments. Evolution has, of course, shaped these responses for the plants!

A final note before continuing: The remaining chapters of this book describe ecological aspects of photosynthetic responses to various environmental conditions such as irradiance or exposure to air in the intertidal, etc., and those responses should therefore preferably be measured *in situ*, i.e. at the site where the plants grow. It is, however, hard to measure gas exchange of individual plants or phytoplankton assemblages in nature since the plants should be enclosed, and it is rather complicated, as well as incorrect, to measure O_2 or CO_2 concentration changes under conditions that, e.g. constrain the water flow around the plants under such enclosed conditions (see above). Therefore, most of the following photosynthetic responses, especially to irradiance, were measured by PAM fluorometry. Again, the advantage of this method is that the plants need not be enclosed, and measurements are very rapid (seconds rather than minutes or even hours when measuring gas exchange). Therefore, many of the examples given in this part will pertain to photosynthesis as measured *in situ* by this method. It must be remembered that PAM fluorometry measures true photosynthesis, and ignores the respiratory component of gas exchange; the pros and cons of this fact are discussed in Section 8.5.

9.2 Light responses of cyanobacteria and microalgae

While some marine microalgae occupy fairly fixed positions as biofilms or as part of the benthic flora (and in that situation are still, like macroalgae and seagrasses, subject to changes in irradiance associated with seasonal, diel and tidal cycles), most are planktonic. While some phytoplankton can maintain their position in the water column through buoyancy regulation or by use of their flagella for vertical movement, others are potentially subject to very large fluctuations in irradiance (and spectral composition) as they circulate around the upper mixed layers of the oceans. Certain groups of algae can cope with such variability in light environment very well; diatoms for example are extremely adaptable and can cope with the high light close to the surface and also function well at very low light as they are transported deep into the mixed layer by water movements. Dinoflagellates on the other hand prefer more stable water columns and, in general, show a preference for lower light levels.

Mechanisms of acclimation by cyanobacteria and eukaryotic microalgae to changing light levels follow the basic strategies outlined above. The tiny picoplanktonic cyanobacterium *Prochlorococcus* seems especially well adapted to the low irradiances and changes in pigmentation allow them to make efficient use of the 'blue-rich' spectral quality of deep ocean environments (far better than other oceanic picoplanktonic cyanobacteria such as *Synechococcus*), and most strains are sensitive to the high light levels found at the surface. There may be different ecotypes of *Prochlorococcus* showing adaptation to the different light environments found through the water column. As such, not only are they subject to the same diel, tidal and seasonal changes in irradiance experienced by benthic species, but they are also exposed to more rapid fluctuations in light as they circulate in the water column. Thus, microalgae and cyanobacteria may be exposed to close to full sunlight when they are at the surface, but within a relatively short period may find themselves at depth at very low light, which is OK too for those species (e.g. of diatoms) that can cope well with those fluctuations.

9.3   Light effects on photosymbionts

Many hermatypic (reef-forming) corals host within their bodies eukaryotic microalgae of the dinoflagellate genus *Symbiodinium*, also commonly called zooxanthellae; these photosymbionts can partly or fully supply them with energy in the form of sugars and possibly other photosynthates too (see Sections 2.3 and 7.2) so that they in effect become autotrophic. Also, many marine sponges host photosymbionts, some as zooxanthellae and some as cyanobacteria, but their role as energy-suppliers is less probable given that the bodies of sponges are much thicker than those of corals and that their photosymbionts for reasons of light capture reside only in the very outer parts of those bodies. On the other hand, they may supply them with essential amino acids, some of which, i.e. the so-called mycosporine-like amino acids or MAAs protect the sponge from harmful UV radiation. Also, it is known that some cyanobacteria produce toxic substances, and if not toxic to the host they may deter other animals from eating the sponge.

Just like for most other plants, the photosynthetic response of photosymbiotic cyanobacteria and zooxanthellae depends on the ambient irradiance in the habitats where they grow naturally. This is exemplified for sponges in Figure 9.5; the cyanobacteria residing at the surface of this sponge species react to irradiance according to where the individual sponges grew. As can be seen, increasing light for growth results in higher light-saturation points and higher maximal photosynthetic rates at light saturation, probably for the same general reasons given for such responses in Section 9.1. Similar responses are found also for sponges hosting zooxanthellae (Figure 9.6). Relative electron transport rates (rel ETR) as a

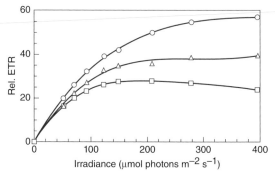

Figure 9.5   Photosynthetic responses of cyanobacteria residing on the surface **of the sponge** ***Theonella swinhoei*** growing at mid-day irradiances of 45 (circles), 30 (triangles) and 23 (squares) μmol photons m^{-2} s^{-1}. Photosynthetic rates on the y-axis are given as relative electron transport rates (relETR) as derived from PAM fluorometric measurements. Reprinted from: Beer S, Ilan M, 1998. In situ measurements of photosynthetic irradiance responses in two marine Red Sea sponges growing under dim light condition. Marine Biology 131: 613–617. Copyright (2013), with permission from Springer.

relative measure of photosynthetic rate is here calculated as the quantum yield (Y, i.e. photosynthetic electron transport per mol photons absorbed) multiplied with the photon flux (i.e. incident irradiance) (see Section 8.3). Usually, Y decreases strongly with irradiance, but not in the case of the zooxanthellae of this sponge. Therefore, it can be concluded that it is irradiance per se, more than shifts in Y with irradiance, that determine the photosynthetic rate of this sponge as a response to irradiance.

The most studied organisms that contain photosymbionts are the corals. Just like sponges, the photosynthetic response of zooxanthellae that inhabit corals depend on the irradiance at which the latter grow. Such irradiances vary not only with depth and season, but also within a coral: in, e.g. a spherical brain coral the irradiance on its lower side may be more than 20 times lower than on its top (see Figure 9.7a). The photosynthetic responses

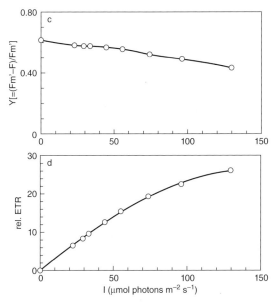

Figure 9.6   Quantum yield (*Y*, panel c) **and photosynthetic rate** (rel ETR, panel d, measured *in situ* by PAM fluorometry) **of the zooxanthellae-containing sponge *Cliona vastifica*** as a function of irradiance. This sponge grew at a midday irradiance of 30 µmol photons m^{-2} s^{-1}. Reprinted from: Beer S, Ilan M, 1998. In situ measurements of photosynthetic irradiance responses in two marine Red Sea sponges growing under dim light condition. Marine Biology 131: 613–617. Copyright (2013), with permission from Springer.

to irradiance of zooxanthellae inhabiting such a spherical, sometimes called 'brain', coral is depicted in Figure 9.7b. While the photosynthetic responses to irradiance are in accord with the light environment that a certain part of the coral was exposed to in nature, we note that the light-saturation point is high throughout, and especially in that part of the coral exposed to high irradiances, albeit only 25% of full sunlight during midday. The fact that such organisms can utilise even the high irradiance of full sunlight (2000 µmol photons m^{-2} s^{-1}) to saturate their photosynthetic production is exceptional; photosynthesis of even high-light

growing terrestrial plants usually saturates well below the irradiance of full sunlight. The question of why corals behave in this way is now open for further investigations!

The three RLCs of Figure 9.7b show the different short-term photosynthetic responses to irradiance of zooxanthellae living under different irradiances. In other words, they show the photosynthetic potential of those zooxanthellae. However, these curves can also be used to predict the degree of acclimation or adaptation to the actual irradiance where they grow. In order to do so, we must look at the photosynthetic response under closer to natural conditions along the *x*-axis. Doing so, we realise that the fact that the low-light zooxanthellae reach only a fraction of the P_{max} than do the high-light ones is of no relevance since they never experience high irradiances. Similarly, the fact that α is much lower in the high-light zooxanthellae has no greater significance since these algae usually see much higher irradiances than those low ones where α is of importance. Thus, according to these RLCs, we can conclude that all populations of zooxanthellae in this coral have adapted to the light environment they usually see during the day: high-light ones feature high photosynthetic rates at the high irradiances they experience by 'investing' in a high capacity downstream of light capture (e.g. in a high level of Rubisco) while low-light ones 'invest' in the capacity to feature high photosynthetic rates at low irradiances (e.g. in high pigment concentrations that can optimise light capture). The zooxanthellae growing in intermediate irradiances show an intermediate response.

The above responses to irradiance were measured as RLCs during a certain season (summer) and at midday for corals growing at a certain depth (3 m). We will now examine if, and how, corals growing at different irradiances along a depth gradient and

(a)   (b)

Figure 9.7   Photosynthetic (ETR) responses to irradiance (I) of zooxanthellae inhabiting the coral *Platy-gyra lamellina* growing at 3 m depth. (a) This spherical brain coral received different irradiances, the midday values of which are noted in panel a (the irradiances values are in μmol photons m^{-2} s^{-1}) and indicated with downward arrows in panel b. (b) The less light that a certain part of the coral received, the lower the light-saturation points (I_{sat}, upward arrows) and maximal photosynthetic rates at light saturation (P_{max}), and the higher the maximal photosynthetic efficiency (α, moved from the origin for clarity). Note that while the high-light zooxanthellae perform at higher rates in high light, the low-light ones show a relatively higher rate at low irradiances. Measured by PAM fluorometry; the three rapid light curves (RLC) were generated using the Diving-PAM. While ETR was calculated using absorption factors (AF), ETR is given as a function of incident light – a fact that we would have changed to absorbed light had we done the experiment today (see 'Alpha, "uses and misuses"' in Section 8.3. Photo with permission from, and thanks to, Katrin Österlund. (a); (b) Reprinted from: Beer S, Ilan M, Eshel A, Weil A, Brickner I, 1998. The use of pulse amplitude modulated (PAM) fluorometry for in situ measurements of photosynthesis in two Red Sea Faviid corals. Marine Biology 131: 607–612. Copyright (2013), with permission from Springer.

throughout seasons, acclimate or adapt[1] to their changing environments during the day. Figure 9.8 shows the photosynthetic response of the branching coral *Stylophora pistillata* to diurnally changing irradiances. When growing under relatively high irradiances, photosynthetic rates before midday are significantly higher than during the afternoon at the same irradiances (Figure 9.8a); i.e. the zooxanthellae experience photoinhibition in the later part

of the day. This photoinhibition is most likely due to the high-light induced damage to the protein complex of PSII termed D1 around midday, such that lower rates of photosynthesis are performed in the afternoon. More exactly, since the photosynthetic rates (relative ETR) were measured here by PAM fluorometry as the product of effective quantum yield (Y, defined as $\Delta F/F_m{}'$) and irradiance, the decreased rate at the same irradiance in the afternoon as before noon is due to a decrease in $\Delta F/F_m{}'$ (see Sections 8.3 and 8.4 for PAM fluorometry and the terms applied for the method). This photoinhibition is remedied already the same evening as evidenced by $\Delta F/F_m{}'$ (and relETR) increasing to the same

[1] We use the term acclimation as the potential of adjusting phenotypically to changing environments; a high degree of acclimation (or acclimatization) shows that an organism has a high plasticity for such adjustments. Adaptation, on the other hand, is a long-term way of adjusting to a certain environment, involving genetic change.

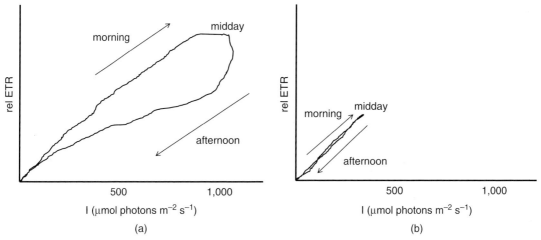

Figure 9.8   Photosynthetic rates as a function of irradiance before (morning) and after (afternoon) midday **of the coral *Stylophora pistillata*** growing at 2 m (a) and 11 m (b) depth. Measured by PAM-fluorometric 'point measurements' (see Section 8.5), i.e. NOT as RLCs. Adapted after: Winters G, Loya Y, Röttgers R, Beer S, 2003. Photoinhibition in shallow water colonies of the coral *Stylophora pistillata* as measured in situ. Limnology and Oceanography 48: 1388–1393. Copyright (2013) by the Association for the Sciences of Limnology and Oceanography, Inc.

values as the following morning. (Another, or a parallel, explanation of the lower rates of photosynthesis in the afternoon is the activation of non-photochemical quenching of photosynthesis, a phenomenon that was explained in Section 8.3.) This type of daily photoinhibition is termed dynamic since it is cured on a daily basis, probably as the D1 protein is re-synthesised. No such diurnal photoinhibition is observed for the zooxanthellae of deeper-growing corals (Figure 9.8b): The midday irradiances were of course lower at depth, and were not saturating for photosynthesis during midday. Rather, and since relETR was the same at equal irradiances during the morning and afternoon, $\Delta F/F_m'$ showed no significant change during the day, and it is irradiance only that determines the rate of photosynthesis. Another measure of photosynthetic efficiency, measured during night-time where the effective quantum yield $\Delta F/F_m'$ is not affected by light, and therefore also called the maximal quantum yield (F_v/F_m,

again see Section 8.3 for terms derived from PAM fluorometry), was constantly lower in the zooxanthellae of the shallow- than in the deep-growing corals, showing another type of photoinhibition that is more chronic, i.e. it persists throughout longer time periods than the diurnally fluctuating dynamic photoinhibition described for the shallow-growing corals. In summary, shallow-growing corals as exemplified here by the common species *Stylophora pistillata* show two types of photoinhibition: a dynamic type that remedies itself at the end of each day and a more chronic type that persists over longer time periods.

One aspect of coral responses to the environment that has been discussed in hundreds of papers is the observation of global coral bleaching. Bleaching of corals occurs when they expel their zooxanthellae to the surrounding water, after which they either die or acquire new zooxanthellae of other types (or clades) that are better adapted to the changes in the environment that caused the bleaching. The most common cause given for coral bleaching is increasing seawater temperatures caused by global warming, but there may be other alternatives such as high irradiances that at least augment possible high-temperature effects. Whatever the cause(s), there is a logical notion that coral bleaching is preceded by a decrease in the effective ($\Delta F/F_m'$) and/or maximal ($\Delta F/F_m$) quantum yields of the zooxanthellae, the latter measures of which thus could be used to predict bleaching events. When doing so, however, it is important to also take into account vari-

ations in photosynthetic traits that are associated with natural, e.g. annual, changes in, e.g. temperature and irradiance. In the example given here, using again the common coral *Stylophora pistillata*, it was found that long-term variations in irradiance affected F_v/F_m more than changes in seawater temperature (see Figure 9.9). The study was able to differentiate between these two abiotic variables (which usually co-vary) because yearly temperature maxima and minima at the site used lagged behind those of irradiance by 3–4 months

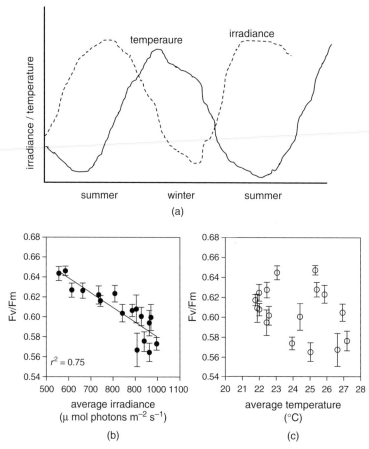

Figure 9.9   Annual variations in temperature (full line) **and irradiance** (broken line) in the northern Red Sea (29°30′N 34°55′E) (a) and annually variations in F_v/F_m values of the coral *Stylophora pistillata* growing at 5 m depth as a function of irradiance (b) and temperature (c) throughout the year. Adapted after: Winters G, Loya Y, Beer S, 2006. In situ measured seasonal fluctuations in Fv/Fm of two common Red Sea corals. Coral Reefs 25: 593–598. Copyright (2013), with permission from Springer.

(Figure 9.9a), enabling assessments of their effects separately. As can be seen, there was a negative correlation between F_v/F_m and irradiance throughout the year (Figure 9.9b) while no such correlation was found with temperature (Figure 9.9c). Whether to call the decrease in F_v/F_m over the year dynamic or chronic photoinhibition is a matter of taste, but may depend on if the same type of zooxanthellae are present in the summer and the winter or not; the former is likely, so just like the diurnal variations in irradiance caused a daily dynamic photoinhibition in this coral, so did its zooxanthellae also on a yearly basis. Based on those results, we suggest that such natural variations should be taken into account when applying PAM fluorometry or other techniques for assessing environmental stressors placed upon corals, some of which could lead to bleaching or other 'catastrophes'.

9.4 Adaptations of Carbon acquisition mechanisms to light

Active Ci acquisition mechanisms, whether based on localised active H^+ extrusion and acidification and enhanced CO_2 supply, or on active transport of HCO_3^-, are all energy requiring. As a consequence it is not surprising that the CCM activity is decreased at lower light levels and the $K_{0.5}$ for CO_2 is increased when the PAR is decreased, i.e. CCM activity is down-regulated. Evidence for down-regulation of CCMs also comes in a few cases from direct measurements of CO_2 accumulation, and more frequently from the observation that algae grown in lower PAR show increased fractionation of stable C isotopes relative to the source inorganic C (more negative $\Delta^{13}C$ values) than do algae grown at high PAR. Such differences are also found for indi-

viduals of the same benthic species that are growing at different depths, and hence PAR. In some florideophyte red macroalgae from sub-tidal or low intertidal environments, and for the intertidal *Lomentaria* growing on the shaded side of rocks, light is presumed to be limiting and the capacity to express CCMs is missing. These adaptational aspects have been detailed by Raven, Ball, Beardall, Giordano and Maberly in their 2005 paper.

9.5 Acclimations of seagrasses to high and low irradiances

Just like macroalgae, the marine angiosperms (or seagrasses) have commonly been considered as 'low-light' plants. This is reflected in their photosynthetic responses to irradiance, which often resembles those low-light plants as illustrated in Figure 9.1. However, many seagrasses live in the intertidal and are therefore, especially in the tropics, subjected to as high irradiances during midday as are terrestrial plants growing on dry land in the same regions. Thus, and indeed, a whole spectrum of light-responses can be found in seagrasses, and those are often in co-ordinance with the average daily irradiances where they grow.

Except for featuring non-photochemical quenching as a way to disseminate high radiation energies (see Section 8.3), at least one genus, *Halophila*, shows another interesting adaptational feature to high irradiances: **chloroplast clumping** (Figure 9.10). In the late 1970s the marine biologist Edward Drew observed that the leaves of the tropical seagrass *Halophila stipulacea* from the intertidal looked paler during midday than during other parts of the day. In a 1979 paper he subsequently ascribed this paleness to chloroplast clumping: the chloroplasts of this seagrass moved

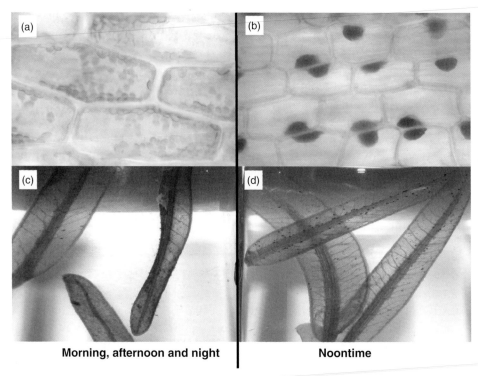

Morning, afternoon and night   **Noontime**

Figure 9.10   Microscope pictures (a, b) and whole-leaf pictures (c, d) of *Halophila stipulacea* leaves growing at 500 μmol photons m^{-2} s^{-1} during morning, afternoon and night-time (a, c) and around midday (b, d) when **chloroplast clumping** is maximal. The leaves of both states (with chloroplasts clumped, d, and dispersed, c) contain the same concentration of chlorophyll. Photos by Yoni Sharon.

to form a tight cluster in the cytoplasm such that most of the cells, and thus the whole leaf, became almost transparent. In the evening, and until the next morning, the chloroplasts moved away from one another and, accordingly, the leaf became darker green and less transparent.

The phenomenon of chloroplast clumping, and its relevance to photosynthetic measurements by PAM fluorometry, photoprotection against high irradiances and UV radiation, was subsequently studied in the PhD work of Yoni Sharon (see also Yoni's Box 9.1). His (and co-workers) *Halophila*-story goes like this: The seagrass *Halophila stipulacea* grows from extremely high irradiances of >2000 μmol photons m^{-2} s^{-1} in tropical intertidal areas to very low ones at depth (<100 μmol pho-

tons m^{-2} s^{-1}). This is true also in the Red Sea, where this plant growth prolifically to depths of >50 m. At high daily irradiances, i.e. at shallow depths, the chloroplasts clump together increasingly till noontime, and then disperse during the afternoon and stay that way throughout the night (see (b) and (a), respectively, in Figure 9.10). Because of this clumping, the entire leaf becomes pale (compare (d) with (c) in that same figure); the concentration of chlorophyll is the same over the day and in the night, and the colour of the leaf is thus purely a function of the degree of chloroplast clumping. Also, these high-light plants show an increasing degree of non-photochemical quenching (NPQ, see Section 8.3) till midday. No such clumping, or expression of NPQ, occurs in lower-light growing plants.

The function of chloroplast clumping in *Halophila stipulacea* appears to be protection of the chloroplasts from high irradiances. Thus, a few peripheral chloroplasts 'sacrifice' themselves for the good of many others within the clump that will be exposed to lower irradiances. This kind of protection from high light is in addition to the protective role of NPQ, and in all the photo-protective features of shallow-growing individuals confers to them a photosynthetic response typical of high-light plants, while those that grow at depth feature photoinhibition at high irradiances (see Figure 9.11). Chloroplast clumping seems so far to be unique to *Halophila stipulacea*. However, the *Halophila*s are all thin-leaved species, and for them to be able to grow at the high irradiances of the tropical intertidal, it is possible that all other 10 or so species feature midday chloroplast clumping. The first one to show this in other species will be honoured with being the first one showing it in other species!

While water is an effective filter of UV radiation (UVR)[2], many marine organisms are sensitive to UVR and have devised ways to protect themselves against this harmful radiation. These ways include the production of UV-filtering compounds called mycosporine-like amino acids (MAAs), which is common also in seagrasses, and the question arose as to the possible degree of protection against UVR by chloroplast clumping. For this, *Halophila stipulacea* plants were grown under high and low irradiances, with and without UV-filtering screens. It was found that clumping occurred under high irradiances only in the presence of UVR. Thus, it was concluded that midday chloroplast clumping in this plant mitigated

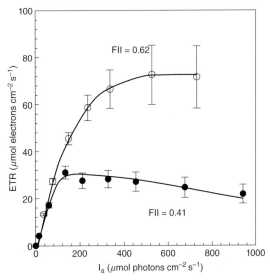

Figure 9.11   Photosynthetic responses to light of the seagrass *Halophila stipulacea* from 1 m (open circles) and ~50 m (closed circles) depth. Photosynthesis was measured as electron transport rates (ETR) as a function of absorbed irradiance (I_a) using PAM fluorometry; ETR was calculated taking into account the different FII factors (see text) for shallow- and deep-growing plants (termed FII in this figure). Reprinted from: Sharon Y, Levitan O, Spungin D, Berman-Frank I, Beer S, 2011. Photoacclimations of the seagrass *Halophila stipulacea* to the dim irradiance at its 48-m depth limit. Limnology and Oceanography 56: 357–362. Copyright (2013) by the Association for the Sciences of Limnology and Oceanography, Inc.

the harmful effects of UVR at least on these plastids.

Halophila stipulacea is an opportunistic seagrass that can photosynthesise and grow under a wide range of irradiances; it can also acclimate quickly to changes in irradiance (see Box 9.1). The leaves are mostly only two cell-layers thick, forming a double-epidermal layer in which each cell contains chloroplasts. Thus, a bit like the highly productive green alga *Ulva*, this seagrass also photosynthesises at high rates (per surface area) and grows quickly under optimal light conditions, and survives under extreme light conditions. At a depth of 50 m, where irradiances are <1 order of magnitude lower than

[2]The attenuation of UVR, like for PAR, depends on the water quality. Strong attenuation of UVR by seawater is the dogma, but in the clear waters of, e.g. the northern Red Sea (where SB and his team has worked a lot) >50% of the surface UVR (280–400 nm) remains at a depth of 5 m; no wonder one's balding scalp gets sunburned when diving there!

at the surface, the plants contain some 3 times more chlorophyll than shallow-growing plants, thus optimising the capture of the relatively few photons reaching that depth. Also the ratio of PSII to PSI (FII) changes in favour of the latter, but the advantage of this is not apparent: The spectrum does change towards blue-green at that depth (see Section 3.1), and the absorption peak of the reaction centre of PSI is at a lower wavelength than that of PSII (680 *vs.* 700 nm), but the quantitative relevance of this is not clear given that all other pigments are present and would efficiently shuttle the energy of all photons towards both reaction centres. Also, the ratio of chlorophyll *a* to chlorophyll *b* remains the same along the depth gradient, and the possible advantage of chlorophyll *b* in absorbing lower-wavelength photons is not present. Again, it may become a challenge for upcoming marine photosynthesis researchers to find the causal connection between high PSI/PSII in deep-growing *Halophila stipulacea* and its ability to utilise the low-irradiance, blue-green-shifted spectrum, light found at depth.

Box 9.1 Acclimations of *Halophila stipulacea* to irradiance along a depth gradient

By **Yoni Sharon**, MBD Energy Ltd, Melbourne, AU (2yonisharon@gmail.com)

The ability of marine macrophytes to develop mechanisms to utilise and/or cope with different irradiances may be crucial for their functioning at temporally dynamic as well as spatially different surroundings. Regarding submerged marine angiosperms (seagrasses), much information has been acquired regarding their photosynthetic responses to irradiance. However, whether these responses are the result of adaptations (long-term selection processes resulting in eco-types) or acclimation processes (short-term plastic responses) were largely unknown. In order to answer this question regarding the tropical seagrass *Halophila stipulacea*, we reciprocally transplanted deep- and shallow-growing plants and observed their photosynthetic behaviour, including photosynthetic rates and pigments content along a few weeks period (that involved long hours of diving in the magical coral reefs of the Red Sea; poor us!).

Our works from the Gulf of Aqaba indicate that *Halophila stipulacea* is a highly plastic seagrass that can acclimate to various light environments quickly (usually within 2 weeks): It was found that all photosynthetic parameters measured featured a high degree of such a plasticity towards various irradiances along a depth gradient (see some of them in Figure 9.12) according to classical adaptation strategies of high- and low-light plants (e.g. P_{max}, I_k, and Y_0, see Section 9.1). Further, this plasticity is in line with the thought that specimens of *Halophila stipulacea* growing at various depths in a meadow are not ecotypes, although the maximal irradiance during midday spans at least 1 order of magnitude along its depth-distribution gradient.

The results of the above acclimation study suggest that the plasticity of *Halophila stipulacea*'s response to different irradiance levels allows for its successful acclimation to the different depths as observed throughout the year. This would then explain the dynamic seasonal meadow extensions as based on clonal growth. While the contents of chlorophylls *a+b* also showed plasticity, the response in the plants transplanted from the high- to the low-light environment was slower than in the low- to high-light transplants (see Figure 9.12c), indicating that a longer time is required for the production of new chlorophyll than its loss. The lack of changes in chlorophyll *a:b* ratios

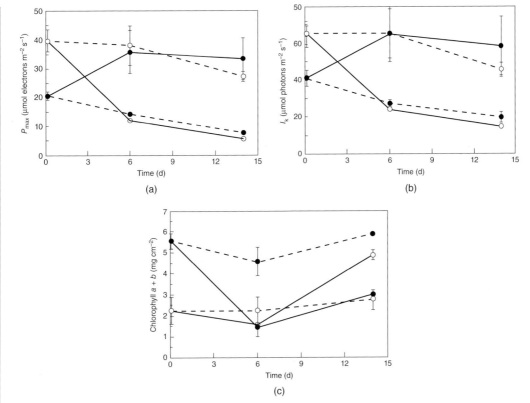

Figure 9.12 Maximal photosynthetic rates (P_{max}, a), onset of light saturation (I_k, b) and chlorophyll $a + b$ contents (c) at 0, 6 and 14 d after transplantation from 8 to 33 m depth (open circles and full lines) and, reciprocally, from 33 to 8 m (filled circles and full line), and the corresponding controls that were not transplanted (dashed lines). Reprinted from: Sharon Y, Silva J, Santos R, Runcie J, Chernihovsky M, Beer S, 2009. Photosynthetic responses of *Halophila stipulacea* to a light gradient: II – Plastic acclimations following transplantations. Aquatic Biology Copyright (2013), with permission from Inter-Research.

in all treatments suggests that this parameter has no role in *in situ* acclimation processes, such as reported also for some other *Halophila* species. While terrestrial shade-adapted plants often increase chlorophyll *b* contents relative to chlorophyll *a*, it may be that the spectral changes with depth prevent such a feature in *Halophila stipulacea*.

   It would be interesting in the future to understand whether *Halophila* is unique in its photosynthetic acclimation process or whether other seagrasses are plastic as well? Is this species depended? Climate dependent? What are the molecular mechanisms enabling this plastic behavior? This will be up to some of you, dear students and young researchers, to find out.

The seagrass *Halophila stipulacea* in Zanzibar. Photo by Mats Björk.

Chapter 10

Photosynthetic acclimations and adaptations to stress in the intertidal

The major stress for organisms and their photosynthetic capacity in the intertidal is desiccation. Therefore, we will first discuss how some intertidal algae and seagrasses cope with desiccation. Other stresses in the intertidal include nutrient stress, salinity stress (as the algae or seagrasses desiccate and salts are concentrated) and the effects of altered water movements that can remove boundary layers and ameliorate desiccation.

10.1 Adaptations of macrophytes to desiccation

Many algae and seagrasses grow in the intertidal and are, accordingly, exposed to air during various parts of the day. On the one hand, this makes them amenable to using atmospheric CO_2, the diffusion rate of which is some 10 000 times higher in air than in water. True, many, or most, algae and seagrasses may not need this since they have a well-developed system for utilising HCO_3^- from seawater as their inorganic carbon (Ci) source for photosynthesis (see Sections 7.3 and 7.4), but some lack efficient HCO_3^- utilisation and may benefit from obtaining CO_2 from the air; in fact some algae and most seagrasses will perform higher rates of photosynthesis just after emergence than when submerged in seawater. The latter is of course true as long as they don't dry out too much; indeed, desiccation is, again, the big drawback when growing in the intertidal, and excessive desiccation will lead to death. When some of the green macroalgae left the seas and formed terrestrial plants some 400 million years ago (the latter of which then 'invaded' Earth), there was a need for measures to evolve that on the one side ensured a water supply to the above-ground parts of the plants (i.e.

Photosynthesis in the Marine Environment, First Edition. Sven Beer, Mats Björk and John Beardall.
© 2014 John Wiley & Sons, Ltd. Published 2014 by John Wiley & Sons, Ltd.
Companion Website: www.wiley.com/go/beer/photosynthesis

roots[1]) and, on the other, hindered the water entering the plants to evaporate (i.e. a water-impermeable cuticle). Macroalgae lack those barriers against losing intracellular water, and are thus more prone to desiccation, the rate of which depends on external factors such as heat and humidity and internal factors such as thallus thickness. As for seagrasses, yes, they have roots but those are mainly used for anchoring the plants to the sediment and for transporting nutrients from the sediments to the leaves; the little water that is transported concomitantly would evaporate quickly during air exposure in the intertidal as these plants lack a cuticle and stomata. Thus, a fine balance may be reached where the rate of desiccation together with the desiccation tolerance of essential processes such as photosynthesis may confer an advantage to some macrophytes in the intertidal, where they thus can have a competitive advantage over others and find a niche for growth. While the mechanisms of desiccation tolerance in macroalgae is not well understood on the cellular level, there are reports on different behaviours of photosynthetic rates to, and the recovery of photosynthesis after, various degrees of desiccation, as exemplified in the following case histories. For seagrasses, there seems to be no physiological adaptation to desiccation, which will be exemplified in the last case history.

10.1.1 The ever-tolerant Ulva

We have used the green alga *Ulva* sp. as a model plant for elucidating the mechanism(s)

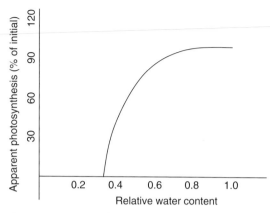

Figure 10.1   Apparent (then called net) **photosynthetic rates of *Ulva* as a function of relative water content.** Photosynthesis was measured as CO_2 exchange in an infrared gas analyser (IRGA) setup (see Section 8.2) and rates are expressed as a percentage of the average rate (100% on the y-axis) during full hydration (1.0 on the x-axis). Adapted after Beer S, Eshel A, 1983. Photosynthesis of *Ulva* sp. I. Effects of desiccation when exposed to air. Journal of Experimental Marine Biology & Ecology 70: 91–97. Copyright (2013), with permission from Elsevier.

of Ci utilisation in macroalgae (in Section 7.3), and also pointed out that this alga can switch its mode of HCO_3^- utilisation according to growth conditions. *Ulva* is found in all the world's oceans, and if something green covers the rocks of the intertidal it is also likely to be *Ulva*. This alga can photosynthesise at full capacity until a relative water content (RWC[2]) of 0.8 (80% water content of the fully hydrated one), where after rates drop but remain positive (i.e. photosynthesis exceeds respiration) until a water content of 35% is reached (Figure 10.1). Comparing this latter value with the

[1] Relative water content (RWC) is the fraction of water remaining in the algal thallus (or seagrass leaf) following various degrees of desiccation. It is calculated after weighing a fully hydrated (FW), and then partly dehydrated (HW), thallus or leaf as RWC = (HW – DW)/(FW – DW) where DW is the dry weight of a fully dehydrated thallus or leaf.

[2] While algae do not possess functional roots (but some species of the green alga *Caulerpa* have rhizoids that anchor them to sediments that are softer than the rocks on which macroalgae usually grow), the seagrasses (or marine angiosperms) do (see Section 2.5). Their role is, however, less (if at all) in supplying water to the shoots and more to anchor them to soft sediments and supply the shoots with nutrients that are plentiful in the sediments.

time it takes *Ulva* thalli to desiccate to such a level of water content, it was concluded that under conditions of, e.g. 25 °C and 30 °C and a relative humidity of 50% this alga could photosynthesise positively for ~90 min and ~30 min, respectively. This, and the relative tolerance of photosynthesis to desiccation, allows *Ulva* to occupy the intertidal successfully.

While the external Ci source of air-exposed *Ulva* in the intertidal is CO_2, we cannot conclude that this Ci form enters the photosynthesising cells. This is because there is always, or at least at the initial stage of air exposure, a film of water just outside the cell membrane, i.e. on top of, or within, the cell wall. It is therefore possible that CO_2 is converted to HCO_3^- on the way through this water film, and especially so if the pH therein is high, as illustrated in Figure 10.2 (see Box 7.2 for considerations of high or low pH in *Ulva*'s diffusion boundary layer). There are thus two potential ways for aerial CO_2 to enter the photosynthesising cells of *Ulva*: 1) CO_2 dissolves in the water film (including the wet cell wall) covering the cells, and then diffuses into the photosynthesising cells where it is fixed and reduced in the chloroplasts; 2) CO_2 dissolves in the water film, and is there converted to HCO_3^-, which is subsequently transported into the photosynthesising cells through an anion-exchange protein (AE) as described in Section 7.3. The latter conversion of CO_2 to HCO_3^- would be aided by an extracellular activity of carbonic anhydrase and a high pH in the water film. Such a high pH is likely, but has not been shown experimentally for air-exposed *Ulva*. This can, however, easily be done by applying a flat-tipped pH electrode to an *Ulva* thallus and then shining a light from below; do it, and you will have a sure publication!

10.1.2 The intertidal Fucus

Fucus is a common genus of brown algae growing in temperate regions. In many areas, three species are distributed along a shallow depth gradient such that *Fucus spiralis* grows highest up, and is thus exposed to more extended periods of desiccation during low tide than *Fucus vesiculosus*, which in turn is more exposed to air than the largely subtidal *Fucus serratus*. While the two latter species desiccate at about the same rate, *Fucus spiralis* desiccates significantly slower, but also rehydrated slower than the other species. When measuring the photosynthetic gas-exchange response to rehydration after desiccation down to 20% (of the original, fully hydrated) water content, the response was approximately the same for all 3 species. However, after desiccating to water contents below 20% there were stark differences in their responses to rehydration such that *Fucus spiralis* could reach almost full photosynthetic

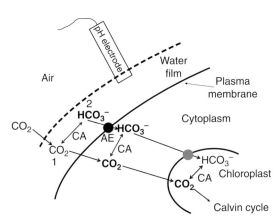

Figure 10.2 Possible **paths of aerial CO_2 in intertidal algae** such as *Ulva* exposed to the atmosphere. Either CO_2 dissolved in the water film covering the thallus diffuses into the cell (path 1), where it can reach the chloroplast as either CO_2 or HCO_3^- (via a hypothetical transporter, grey circle), or CO_2 is converted to HCO_3^- via extracellularly acting carbonic anhydrase (CA) in the water film (path 2). If the latter, then HCO_3^- would enter the cell via an anion-exchange protein in the plasma membrane (AE) and the chloroplast as CO_2 or HCO_3^- just like for path 1. Drawing by Sven Beer.

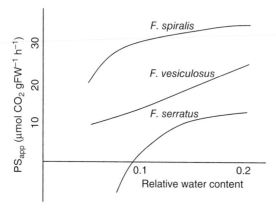

Figure 10.3 The apparent photosynthetic response of three _Fucus_ species (_F. spiralis, F. vesiculosus_ and _F. serratus_, growing in the high, middle and low intertidal, respectively) **to resubmersion after desiccation.** Apparent photosynthetic gas exchange rates (expressed as µmol CO$_2$ g^{-1} fresh weight h^{-1}) after rehydration following desiccation to various water contents, from 20% down to 5–8% (relative to fully hydrated thalli, on the x-axis). Adapted after: Beer S, Kautsky L, 1992. The recovery of net photosynthesis during rehydration of three Fucus species from the Swedish west coast following exposure to air. Botanica Marina 35: 487–491. Copyright (2013), with permission from De Gruyter. (Available at: www.reference-global.com.)

rates after rehydrating from a water content of 5–10% while _Fucus vesiculosus_ (the one occupying the middle zone) could not regain full photosynthetic rates after having lost >20% of its water (Figure 10.3). Most sensitive was _Fucus serratus_, which could not regain even low positive photosynthetic rates after rehydrating from 10% water content (but respired only). Again, the time it takes for algae to reach those critical water contents at which they can no longer photosynthesise, or cannot regain positive photosynthetic gas-exchange rates following rehydration, depends on external factors such as temperature and humidity as well as internal factor such as thallus morphology. For _Fucus_ from a setting on the Swedish west coast (25 °C and 60% relative humidity) it took thalli

1–2 h (depending on species) to reach a water content of 20%, while for the thin-thallied _Ulva_ in a Mediterranean setting (30 °C and 50% humidity) it took only 30 min to reach such a level of desiccation. In summary for _Fucus_, it seems that the photosynthetic performance during rehydration following dehydration of the thalli determines at least partly the zonation pattern of the different species along a gradient of wetness in the intertidal.

10.1.3 The extremely tolerant _Porphyra_

Among the macroalgae, some red species, especially of the genus _Porphyra_, are very resistant to desiccation. This is realised in their photosynthetic systems being able to function at very low cellular water contents and, especially, being able to revive after prolonged periods of desiccation. A case to consider is that of _Porphyra linearis_, an alga that, e.g. grows highest up in the intertidal of the eastern Mediterranean. In air, this alga can photosynthesise at full capacity until a relative water content of 0.6, where after rates drop linearly until a water content of some 10%, at which value there is no longer any measurable gas exchange in the light (Figure 10.4a). When fully dry, this alga may contain about 5% water (probably bound hygroscopically to salts), at which stage both photosynthesis and respiration shuts down. However, after rewetting the thalli after, e.g. 48 h of such dryness, net photosynthesis becomes positive at a water content of 30% and full rates are obtained after the water content is back to some 80% (Figure 10.4b). Not only is the photosynthetic apparatus active down to very low water contents, but it also survives, as does the entire plant, experimental dehydration down to 5% water content for up to 3 weeks, after which photosynthesis becomes

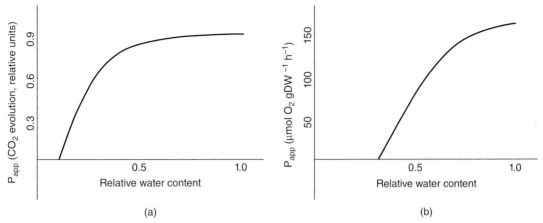

Figure 10.4   **Apparent photosynthetic rates (P_{app}) of *Porphyra linearis*) in air** as a function of relative water content (a), **and** as a function of relative water content **during rehydration** (b) following 48 h of full dehydration (to 5% water content). See text for details. Adapted after: Lipkin Y, Beer S, Eshel A, 1993. The ability of *Porphyra linearis* (Rhodophyta) to tolerate prolonged periods of desiccation. Botanica Marina 36: 517–523. Copyright (2013), with permission from De Gruyter. (Available at: www.reference-global.com.)

active again after rehydration. The latter time fits well with the time that these thalli were observed to remain in dry conditions high up in the intertidal between two consecutive high tides, during which the thalli are naturally rewetted. Incidentally, *Porphyra* species from Japan are also called Nori, and they are used for wrapping the rice in sushi. Given the extreme desiccation resistance of these algae, one should be aware that they may come alive in the mouth of sushi-eaters. However, their temporary rehydration will not be felt as much as if it was the fish in sushi recovering. A similar degree of desiccation tolerance is found in the red algae *Bostrychia* and *Caloglossa*, found frequently on, for instance, mangrove roots. These can also lose water to a state where they will crumble to the touch, but will revive photosynthetic activity soon after rehydration.

The red alga *Porphyra* is morphologically similar to the green alga *Ulva*; both have thin thalli containing only two cell layers (while the brown *Fucus* species have much thicker thalli). These algae possess no barriers against desiccation, and the two thin-thallied ones dry out

at the same rate as does a piece of wetted filter paper. Therefore, it is not the rate of desiccation that determines the vertical position of an alga along a gradient of wetness, but rather the coping of biochemical pathways and, it seems, especially the photosynthetic system to thallus water content.

There are few microalgae and cyanobacteria that live in the intertidal and are, therefore, also exposed to periodic desiccation stress. There have been even fewer studies on how these species cope with desiccation, although there is a large literature on species that form crusts in desert regions and are thus exposed to prolonged periods without water. These though lie outside the scope of this book!

In summary, there seems to be a general correlation between the sensitivity of the photosynthetic apparatus (more than the respiratory one) to desiccation and the occurrence of macroalgae along a vertical gradient in the intertidal: the less sensitive (i.e. the more tolerant), the higher up the algae can grow. This is especially true if the sensitivity to desiccation is measured as a function of the ability to regain photosynthetic rates following rehydration during re-submergence. While this correlation exists, the mechanism of protecting the photosynthetic system against desiccation is largely unknown: Young scientists: go for it!

10.1.4 Acclimations of seagrasses to desiccation (or not)

Just like some macroalgae, many seagrasses grow also in the intertidal. This is true not only for temperate species such as *Zostera marina*, but also for many tropical species that are exposed to the hot air for many hours a day. Here, as in macroalgae, it was assumed that the higher up a species could grow, the more tolerant it would be towards desiccation. However, unlike in macroalgae, this could not be confirmed experimentally. Rather, quite the opposite was found: the higher up in the intertidal zone that a species could grow, the more sensitive its photosynthetic performance was to desiccation, and *vice versa*, the deeper they grew in the zone, the more resistant was their photosynthesis to desiccation. This was culminated by some species that only grew submerged and, contrary to 'logics', showed the highest resistance to desiccation of all. This phenomenon can be exemplified by comparing the species growing highest up (*Halophila ovalis*) and lowest down (*Enhalus acoroides*) along a depth gradient in Zanzibar, Eastern Africa, that includes the intertidal zone (Figure 10.5). As a measure of photosynthetic activity, the effective quantum yield, Y, measured as $\Delta F/F_{m}'$, was assessed during a desiccation cycle (Figures 10.5a and c), and the same parameter was then followed after rehydration following dehydration to certain pre-determined relative water contents (Figures 10.5b and d). As can be seen, *Halophila ovalis*'s extreme sensitivity to desiccation was manifested by a drop in Y to half its original value after losing only 10% of its water

content. In contrast, *Enhalus acoroides* maintained over half of its Y value after losing 25% of its water content. Similarly, the desiccation-sensitive plant could not regain full photosynthetic rates upon re-submergence at dehydration levels of 60% water content while the tolerant one could regain full photosynthetic rates after being desiccated to a water content of 45%. Not only was the sensitivity to desiccation opposite to what would predicted both logically and in analogy with macroalgae (see the preceding section), but the actual time it took the leaves to desiccate was much slower in the subtidal species than in the one growing highest up in the intertidal.

Two principal things can be learned from the above findings: First, nature does not always act in a manner thought as 'logical' to human thinking and, secondly, such 'negative' results can and should be published (as were the results exemplified in Figure 10.5) so that others can look for different causes of, in this case, seagrass zonation patterns. In this case, if the photosynthetic response to desiccation is not in line with the vertical distribution pattern of seagrass species, then what is? This is, again, for future marine scientists to determine. What we (MB and SB) have observed is that the upper growing seagrass species are smaller than those growing lower down, and certainly smaller than *Enhalus acoroides*, that has been named as arguably the world's largest seagrass. Together with the smallness comes a flexible petiole that lets the leaves lie flat on the sand, the latter of which is still moist during low tide. Thus, the 'secret' of *Halophila ovalis* (and other seagrasses that have smallish flexible leaves) is that they are never really exposed to high degrees of desiccation in the intertidal, and so their sensitivity to desiccation is not generally realised. On the other hand, the larger-leaved seagrasses that grow deeper have more sturdy leaves that cannot bend down towards the moist sand, and will thus be exposed directly to the (usually)

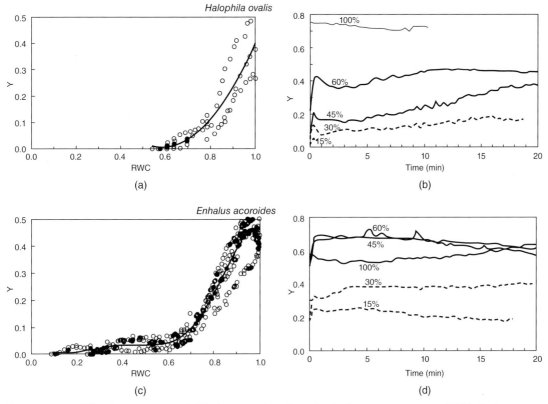

Figure 10.5   Effective quantum yields (*Y*) as a function of relative water content (RWC) during emergence (a and c) and during re-submersion following desiccation to various RWCs (b and d) **for the seagrasses** *Halophila ovalis* (growing highest up in the intertidal) and *Enhalus acoroides* (growing deepest down). *Y* was measured in situ using a Diving-PAM under 500 µmol photons m^{-2} s^{-1}. Reprinted from: Björk M, Uku J, Weil A, Beer S, 1999. Photosynthetic tolerances to desiccation of tropical intertidal seagrasses. Marine Ecology Progress Series 191: 121–126. Copyright (2013), with permission from Inter-Research.

hot air; in this case they will dry out completely and the demonstrated better performance during partial desiccation will become irrelevant.

10.2   Other stresses in the intertidal

Algae inhabiting the intertidal are subject to a range of other environmental stresses that can directly or indirectly influence photosynthesis. These include temperature extremes that will have the same effect on photosynthesis that they do on any biochemical process: Below optimal temperatures, exposure to higher temperature when the tide is out will increase photosynthetic carbon fixation rates according to the Q_{10} rule, i.e. roughly a doubling or tripling of rate with a 10 °C rise in temperature.

Macroalgae gain their nutrients from the water column, so when the tide is out, they cannot obtain these nutrients and the ensuing nutrient deprivation can affect their photosynthetic capacity. Nutrient (mainly N and P) limitation thus causes a decrease in photosynthetic capacity, both in terms of light harvesting (chlorophyll synthesis is especially sensitive to N limitation) and carbon fixation (e.g. Rubisco levels are diminished under N and, to a lesser extent P, limitation). Many algae, however, and especially those from the upper intertidal, feature high-capacity uptake systems that can accumulate and store nutrients internally when the tide is in.

The coastal zone is often a region of high water turbulence, with breaking waves and tidal action. Aside from mechanical stress, water turbulence can have a number of more direct effects on algal photosynthesis: Water spray in the 'splash zone' can help minimise desiccation stress. Turbulent flow over the algal thallus can minimise boundary layer thickness. Given Fick's law of diffusion, in theory a decrease in boundary layer thickness will improve fluxes of inorganic carbon and nutrients to the thallus surface and thereby enhance photosynthetic rates (see Hurd 2000 for a full treatment of the effects of water turbulence). However, evidence for enhanced productivity in areas subject to increased water motion is, at present, poor.

Chapter 11

How some marine plants modify the environment for other organisms

Although not much research has been done yet in this rather new field of marine plant ecophysiology, it is probable that photosynthetic properties of one plant, or a plant community, can have influences on other species in its surroundings. Epiphytism is a common feature among marine plants, and some of the physical effects of epiphytes on their hosts will be treated in Section 11.1. Influences can be either inhibitory to the other plants or can be favourable; these two possibilities will be exemplified for some marine macrophytes in Sections 11.2 and 11.3, respectively. An interesting phenomenon among the 'microalgae' is the fact that some release CO_2 during photosynthesis, and this CO_2 could be used by others – this will be described in Section 11.4.

11.1 Epiphytes and other 'thieves'

If epiphytes or, in some cases, co-occurring plants such as filamentous algae (Figure 2.6d) or, often, the opportunistically growing *Ulva* (see Figure 11.1) cover a marine 'host' plant they will 1) reduce the light that penetrates to the host plant underneath, 2) shift the quality (i.e. spectrum) of the remaining light since the epiphyte will reflect or absorb some wavelengths more than others, and thus a green algal epiphyte will cause the light reaching the plants underneath to be green, 3) compete with the host plant for nutrients, and 4) compete with the host plant for CO_2. These

Photosynthesis in the Marine Environment, First Edition. Sven Beer, Mats Björk and John Beardall.
© 2014 John Wiley & Sons, Ltd. Published 2014 by John Wiley & Sons, Ltd.
Companion Website: www.wiley.com/go/beer/photosynthesis

Figure 11.1 The seagrass *Zostera marina* **partly** covered by a thick layer of finely branched *Ulva*. Photo by Mats Björk.

effects of epiphytes will now be discussed one by one:

1. The reduction of the light can of course have a negative effect on the plants, but only if the plant underneath the epiphyte (i.e. the host) is light limited. In many areas, including the tropics, light levels are higher than is needed for the plant, and here the epiphyte cover could actually be protecting the plants from photoinhibition or photodamage.

2. The shift in spectral distribution of the light that penetrates through an epiphyte can also have an effect on the photosynthetic performance of the plants beneath it since some wavelengths are absorbed better than others. Green light is, e.g. known to be less absorbed by single-algal cells or thin macroalgae (while thicker thalli absorb more of the incident photons of all wave-

lengths, see Section 3.1); indeed, when looking at thicker tissues like the thalli of larger macroalgae (also called fronds) or the multi-cell layers of seagrass leaves, it appears that most light is absorbed; the light that is less absorbed by the photosynthetic pigments in the epidermis will penetrate deeper into the tissue, but will eventually be absorbed and drive photosynthesis as efficiently as other wavelengths. Thus, epiphytes can be seen as part of the outer layer of a macroalga or seagrass through which selected wavelengths will penetrate into the underlying layers and there give rise to photosynthesis of the host. It has even been suggested that this will increase overall performance of the host plant since this way the photons are better distributed in the photosynthetic tissue.

3. Nutrients like N and P are often a limiting factor for marine plant production, and under such circumstances it will be a disadvantage to the macroalgal host if there is a competitor present in the form of an epiphyte that will have more ready access to in seawater dissolved nutrients. (This is less apparent in seagrasses that absorb nutrients also from the sediment.) However, some epiphytes, like some cyanobacteria, fix N_2, which might actually increase the availability of usable N sources to the host plants and, in some areas, this can be a major source of N to the macrophytes.

4. The uptake of CO_2 by photosynthetic epiphytes and fast-growing co-occurring algae like *Ulva* (Figure 11.1), as well as the host, can shift the pH upwards within the diffusion boundary layer and, eventually, the surrounding seawater (see Section 11.3). This will induce changes in carbon 'speciation' towards HCO_3^- and CO_3^{2-} and such plant-generated pH shifts could extend up to pH 10 and, then, have a significant negative effects on the photosynthetic capacity of the two co-occurring organisms (see Figure 11.2).

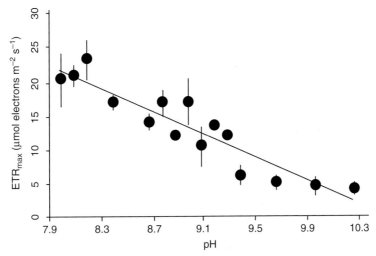

Figure 11.2   The maximal photosynthetic electron transport rate (ETR_{max}) **of *Zostera marina* as a function of pH** (as a proxy for Ci availability). Photosynthesis was measured in a closed system where *Ulva* had depleted the Ci content of the seawater so that pH increased (according to Figure 3.6) so as to simulate field conditions as illustrated also in Figure 11.1. Adapted after: Mvungi EF, Lyimo TJ, Björk M, 2012. When *Zostera marina* is intermixed with *Ulva*, its photosynthesis is reduced by increased pH and lower light, but not by changes in light quality. Aquatic Botany, 102: 44–49. Copyright (2013), with permission from Elsevier.

11.2   *Ulva* can generate its own empires

Some algae seem to occupy their own private niche by preventing other algae from growing in the same habitat. A case exemplifying this can be for *Ulva* sp. growing in rockpools along many shores. Rockpools are rather isolated from the open sea, their water being exchanged infrequently during storm events or extreme high tides. If these pools are occupied by *Ulva*, then usually no other plants grow in them. So what is it with *Ulva* that hinders other algae from growing in its vicinity? First, *Ulva* differs from virtually all other macroalgae by its apparent ability to take up HCO_3^- in exchange for OH^-. In this way, the pH in its vicinity can be >10 during the day. Under such circumstances, other macroalgae that rely on extracellular HCO_3^- to CO_2 conversions could not photosynthesise at significant rates since the CO_2 concentration would be so low

that no inward diffusion could take place (see Section 7.3 for HCO_3^- utilisation mechanisms in macroalgae). Secondly, *Ulva* is an opportunistically growing alga that can establish itself quickly in, e.g. denuded rockpools. Once established, its ability to raise the pH to very high values would exclude other algae from growing there.

While we have seen rockpools occupied by *Ulva* as its sole macroalga in both tropical and temperate areas of the world, some experiments were performed in rockpools along the Swedish west coast in order to establish if its photosynthetic trait of HCO_3^- uptake could be of advantage in its potential competition with other algae. Such a rockpool is depicted in Figure 11.3, and it is occupied by *Ulva intestinalis* only, while many other macroalgae grow in the small bay just beyond it. The pH in such rockpools reaches values of >10 during the day, while the pH of the bay may increase to <9. At the same time, the concentration of inorganic carbon (Ci) decreased from close to

Figure 11.3   A rockpool above a small bay on an island in the archipelago of the Swedish west coast. Observe *Ulva intestinalis* (along the rim) as its sole macroalgal inhabitant (while many other species grow in the small bay). Reprinted from: Björk M, Axelsson L, Beer S, 2004. Why is *Ulva intestinalis* the only macroalga inhabiting isolated rockpools along the Swedish Atlantic coast? Marine Ecology Progress Series. 284: 109–116. Copyright (2013), with permission from Inter-Research.

2 mM in the morning to ~0.6 mM at midday (while it remained ~2 mM in the bay). Two other algal species common in the bay, the brown alga *Fucus vesiculosus* and the red *Chondrus crispus*, were then transferred to the rockpool, and some of their photosynthetic parameters were followed by PAM fluorometry: The midday electron transport rate (ETR) of *Fucus* decreased to 30% of the value in the bay when transferred to the rockpool, and that of *Chondrus* was zero in the rockpool as compared to 15 μmol electrons m^{-2} s^{-1} in the bay (see Figure 11.4). At the same time, *Ulva* reduced its ETR by only 40% in the rockpool as compared to the bay. As an apparent result, *Chondrus* had photo-bleached completely after only one day in the rockpool, and died shortly thereafter, and

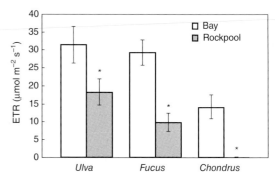

Figure 11.4   Midday photosynthetic electron transport rates (ETR) of *Ulva intestinalis*, *Fucus vesiculosus* and *Chondrus crispus* **in either the small bay or the rockpool** depicted in Figure 11.3. Adapted after: Björk M, Axelsson L, Beer S, 2004. Why is *Ulva intestinalis* the only macroalga inhabiting isolated rockpools along the Swedish Atlantic coast? Marine Ecology Progress Series. 284: 109–116. Copyright (2013), with permission from Inter-Research.

Fucus was on its way to bleach, while *Ulva* was unaffected by the high pH and low Ci concentration. These results were confirmed in simulated rockpools (read: plastic bins) where the three species were kept either alone or together: Figure 11.5 shows that F_v/F_m, which is a general measure of stress affecting photosystem II (see Section 8.5), was significantly lowered in both *Fucus* and *Chondrus* when they were kept together with *Ulva* and, accordingly, were subjected to pH values >10 and low Ci concentrations. Under such conditions, there was apparently likely not a high enough CO_2 concentration generated in the diffusion boundary layers of the brown and red algae in order to support photosynthesis, and so the electrons generated in the light reactions caused photobleaching instead.

As a summary of the above, *Ulva* can, under conditions of high pH that it generates (see also Section 7.3), take up HCO_3^- from the seawater in exchange for OH^-, and can thus photosynthesise at high pH values. This renders it an advantage over most other macroalgae that rely on converting HCO_3^- to CO_2 in their diffusion boundary layers; this process is not effective at

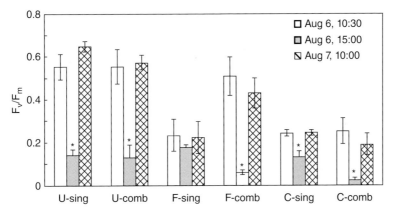

Figure 11.5   Daily **changes in F_v/F_m** (measured after 10 min dark adaptation) in *Ulva intestinalis*, *Fucus vesiculosus* and *Chondrus crispus* grown in **'simulated rockpools'** either alone (sing) or together with the others (comb). Adapted after: Björk M, Axelsson L, Beer S, 2004. Why is *Ulva intestinalis* the only macroalga inhabiting isolated rockpools along the Swedish Atlantic coast? Marine Ecology Progress Series. 284: 109–116. Copyright (2013), with permission from Inter-Research.

high pH values because the CO_2 generated is, in equilibrium with HCO_3^-, at extremely low concentrations. This is an example of how a photosynthetic trait of one algae can cause it to outcompete others with a different way of Ci acquisition. More such examples of competition via photosynthetic traits are surely out there to be explored!

11.3   Seagrasses can alter environments for macroalgae and *vice versa*

Seagrass photosynthesis, like that for all marine plants, can change the pH in the surrounding seawater where they grow. This is true especially for enclosed areas where the diurnal exchange of CO_2 with the atmosphere is slower than the photosynthetic removal of Ci (or the nocturnal addition of CO_2 is faster than its equilibration with the atmosphere). Under such situations, pH increases during the day because Ci, including CO_2 in equilibrium with H_2CO_3 (carbonic acid) in the seawater, decreases. While such a pH increase will slow

down the photosynthetic activity of most algae and seagrasses (for the reasons described in Section 7.3 as well as in the previous section), it may enhance other processes such as calcification (for reasons given in Section 7.5).

One example of how seagrass photosynthesis can enhance algal calcification can be given from a tropical bay in Zanzibar, East Africa (see Figure 11.6a). Here, most of the sand is remnants from the calcifying green alga *Halimeda renschii*. This alga grows among several tropical species (see Figure 11.6b) including *Cymodocea rotundata* (pictured), *Thalassia hemprichii* (perhaps mainly) and *Enhalus acoroides* (pictured in Figure 11.6a)). These seagrasses were all previously shown to appear in high-pH waters generated by their own photosynthetic activity. Now, it is known that calcification rates of calcifying algae is enhanced by high-pH surroundings (see, e.g. the reference of Semesi et al., 2009), and the question asked was if the photosynthetic performance of seagrasses could enhance the calcification rate of calcifying algae growing within a seagrass meadow. For this purpose, experiments were set up within the meadow during low tide in which calcifying algae were kept alone

Figure 11.6 Chwaka Bay, Zanzibar, Tanzania: a) The seagrass *Enhalus acoroides* is seen in the foreground, and the sand between the seagrass patches is the CaCO₃-remainders of the calcifying alga *Halimeda renschii*; b) The **calcifying green alga** *Halimeda renschii* growing together with the **seagrass** *Cymodocea rotundata*. Photos by Sven Beer (a) and Mats Björk (b).

or together with seagrasses in enclosures open only to the atmosphere. It was found that pH indeed increased to high values in all enclosures where seagrasses were present, and that the calcification rate of 3 different calcifying algae increased significantly when inhabiting such settings. One example is shown in Figure 11.7, where the green alga *Halimeda renschii* can be seen to approximately double its rate of calcification when in the presence of

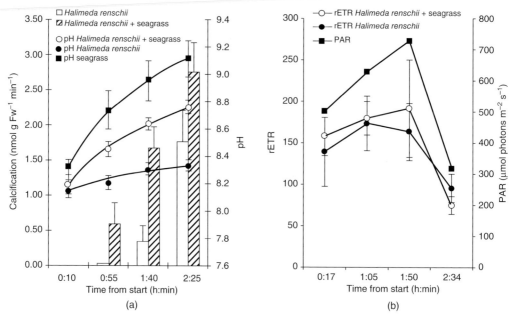

Figure 11.7 **Calcification rate and pH** (a), and relative electron transport rate (rETR, calculated as Y•PAR) and irradiance (PAR) along the incubation time of the **alga** *Halimeda renschii* alone, the **seagrass** *Thalassia hemprichii* alone, or the **two together** (b). Adapted from: Semesi, S., Beer, S. and Björk, M. 2009. Seagrass photosynthesis controls rates of calcification and photosynthesis of calcareous macroalgae in a tropical seagrass meadow. Marine Ecology Progress Series. 382: 41–47. Copyright (2013), with permission from Inter-Research.

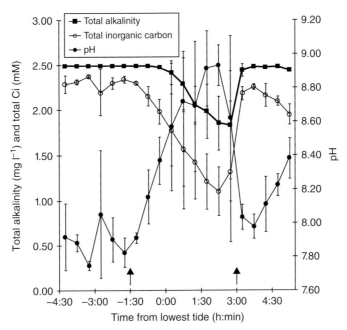

Figure 11.8   Daily changes in pH, alkalinity (the reciprocal of which is a measure of calcification) **and inorganic carbon** during part of a tidal cycle in the intertidal of Chwaka Bay, Zanzibar, Tanzania, inhabited mainly by calcifying algae and seagrasses. The left arrow indicates the beginning of the incoming tide and the right arrow the beginning of the outgoing tide. The water depth between the arrows was <0.5 m. Adapted from: Semesi, S., Beer, S. and Björk, M. 2009. Seagrass photosynthesis controls rates of calcification and photosynthesis of calcareous macroalgae in a tropical seagrass meadow. Marine Ecology Progress Series. 382: 41–47. Copyright (2013), with permission from Inter-Research.

seagrasses, the latter of which increased the seawater pH to ~9.

In the bay where the above experiment took place, pH rises from 7.9 to 8.8 during a typical low-tide period of some 1.5 h (Figure 11.8). This is due to the photosynthetic activity of the plants growing in the bay as evidenced by the concomitant decrease in Ci concentration. Also, the total alkalinity decreased, which is taken as evidence that calcification took place, as discussed earlier, photosynthesis only will lower the Ci concentration while keeping alkalinity unchanged (see Section 3.3). This is another example of how photosynthesis of one marine plant, or a group of plants (seagrasses) can influence the performance of another group (in this case calcifying macroalgae), and again we say that this is just the beginning, and other interactions based on the direct outcome of photosynthetic parameters of one plant can be the basis of many other interactions with other plants.

11.4   Cyanobacteria and eukaryotic microalgae

In phytoplankton populations, there is frequently a clear succession of species during the course of the year. While this is frequently driven by seasonal changes in light availability and temperature, it is also possible that competition between species is driven by changes associated with alterations in the immediate environment effected by the algae themselves. For instance, coastal populations can be dominated by diatoms until the Si in the water column is depleted to a level where diatom

---

Box 11.1   Some phytoplankton excrete inorganic carbon during photosynthesis

By **Aaron Kaplan,** The Hebrew University, IL   (aaronka@vms.huji.ac.il)

Our research group's analyses of the inorganic carbon (Ci) fluxes between phytoplankton organisms representing various phyla and their surroundings indicated that: a) the fluxes are enormous, in some cases several fold higher than the photosynthetic rate; b) some organisms take up HCO_3^- and evolve CO_2, others do the reverse, and yet others, such as the coccolithophore *Emiliania huxleyi*, can change the direction depending on the irradiance; c) in many cases the fluxes are masked by the activity of carbonic anhydrase (CA) that facilitate chemical equilibria between the Ci species in the medium and thereby disguise the Ci fluxes that otherwise shift the system away from equilibrium.

In addition to its importance as an essential part of the functioning of the CCM and the physiological consequents thereof, discussed in Section 7.1, and to the dissipation of excess light energy and pH homeostasis, there is another intriguing possibility that the Ci cycling by some organisms may help supply CO_2 to others in the water body. As an example, the enormous CO_2 evolution observed in the case of *Synechococcus* sp. WH7803, 8-fold larger than CO_2 fixation in photosynthesis, may help other phytoplankton located in its close proximity to acquire the CO_2 produced. Such associations may stimulate productivity in the marine and fresh water bodies. However, it is difficult to experimentally assess the magnitude and the ecological significance of such phenomenon for a simple reason: The physicochemical processes in the medium tend to convert the CO_2 evolved back to HCO_3^- and this is further stimulated by CA activity in the surrounding media or the cells' periplasmic space. Consequently, to show such fluxes experimentally it is necessary to use cell densities well above those normally present in the water bodies (perhaps with the exception of massive algal or cyanobacterial blooms) or apply CA inhibitors. Nevertheless, a sophisticated approach employing stable C and O isotopes may help to assess the importance, if any, of such processes.

The following papers are instrumental to our above discoveries, and are recommended to the interested reader: Tchernov D, Hassidim M, Luz B, Sukenik A, Reinhold L, Kaplan A, 1997. Sustained net CO_2 evolution during photosynthesis by marine microorganisms. Current Biology 7: 723–728 and Tchernov D, Silverman J, Luz B, Reinhold L, Kaplan A, 2003. Massive light-dependent cycling of inorganic carbon between photosynthetic microorganisms and their surroundings. Photosynthesis Research 77: 95–103.

---

growth is severely limited and other, non-silicifying, species can gain an advantage. In the, brackish-water, Gippsland Lakes system in Victoria, Australia, intense diatom or dinoflagellate growth depletes N in the water column to such a degree that populations of these groups crash, the decay of their organic matter leading to anoxic conditions in, and P release from, the sediment. This creates ideal conditions for the N_2-fixing toxic cyanobacterium *Nodularia spumigenea* that can, given the right conditions of temperature and salinity, become dominant and form extensive blooms. Another intriguing possibility is that activities of carbon acquisition systems of some species may help to provide Ci, in one form or another, for others. This is discussed by Aaron Kaplan in the box above.

Chapter 12

Future perspectives on marine photosynthesis

12.1 'Harvesting' marine plant photosynthesis

For centuries, humans have been harvesting algal biomass for a range of products. In Asia particularly there is a rich history going back millennia of macroalgal cultivation for food, though people in other countries have also harvested algal biomass for food and specific products such as alginates and carrageenans, which also have industrial applications. In previous centuries, macroalgae were harvested as sources of potash and iodine.

Industrial applications of alginates continues to the present day with uses in the food, cosmetics, medical and chemical industries. Carrageenans still command a market, as phycocolloids, of >US$ 600 million (2007/2008 figures) with alginate having worth about half this and agar having a value a quarter of that for carrageenans. Not bad for such special products of, ultimately, the marine photosynthetic process!

More recently there has been increasing interest in using algae as a source of biomass for a range of nutraceutical products (food supplements that provide 'extra' health benefits) such as omega-3 fatty acids, antioxidants and an array of bioactive substances of potential application in medicine. Interest in this area is mostly centred around growth of microalgae. Microalgal culture has also been strongly advocated as a source of oils for use in biofuels as algae can, unlike alternatives such as the palms grown for oil, be grown in areas unsuitable for conventional agriculture. While there have been some excessively optimistic claims about the potential for algal biofuels and their economic feasibility, it is clear that some microalgae are capable of diverting a large proportion of their photosynthetic product into lipids, which can then be converted to biofuels and other useful products (see Williams and Laurens 2010 for a critical review).

There is also considerable interest in using photosynthesis in algal production plants to strip out CO_2 from industrial operations such

Photosynthesis in the Marine Environment, First Edition. Sven Beer, Mats Björk and John Beardall.
© 2014 John Wiley & Sons, Ltd. Published 2014 by John Wiley & Sons, Ltd.
Companion Website: www.wiley.com/go/beer/photosynthesis

as power plants. Thus, algal culture would use photosynthesis to ameliorate anthropogenic CO_2 emissions. In theory this is a laudable objective, though there are still significant issues to address in setting up such a system; these include the lower pH associated with a gas stream containing 10–12% CO_2, contaminants, especially sulphur compounds, and the energy economy of harvesting and de-watering of the algal biomass (see again Williams and Laurens 2010).

The oceans and the algae and seagrasses living in them have been responsible for sequestering ~40% of the anthropogenically released CO_2 since the onset of the Industrial Revolution. The drawdown of CO_2 and its sequestration in deep-ocean waters and sediments, as well as carbon locked up in mangrove ecosystems, saltmarshes and the rhizosphere of certain seagrasses, is referred to as 'Blue Carbon'. Although algal photosynthetic products are not currently recognised as Blue Carbon, if the carbon assimilated is locked up and not immediately released back into the atmosphere as CO_2 from eaten (respired) or from burnt biofuel material it does represent a net carbon removal from the atmosphere. Even if algal-derived biofuels are burnt to release energy, with release of CO_2, they do provide a way of by-passing the use of fossil fuels, thereby ameliorating global warming.

In order to boost algal productivity in carbon remediation trials, there have also been attempts to carry out large-scale ocean fertilization experiments with Fe or N enrichments in the open ocean. The concept here is that fertilization will enhance primary productivity of phytoplankton, which will then sink to the deep ocean, locking away the carbon for centuries. Such long-term sequestration may be eligible for 'carbon credits'. However, the amount of net drawdown into the deep ocean from Fe-enrichment in the Southern Ocean for example is very small and unlikely to make a significant contribution to global CO_2 sequestration. Furthermore, nutrient enrichment raises the possibility of bringing about unwanted changes in ecosystems – for example plans to enrich the Sulu Sea with urea have been criticised because high levels of urea in seawater have been linked to toxic dinoflagellate blooms (Glibert and co-workers, 2008). We mess with the oceans at our peril!

12.2   Predictions for the future

Our planet is currently going through changes, the pace of which is unparalleled in geological history. Current predictions for the A1F1 scenario (predicting a very rapid economic growth and a high usage of fossil fuels) of the Intergovernmental Panel on Climate Change (IPCC) are for an increase in atmospheric CO_2 to ~1000 ppm and a 4 °C rise in global average temperature by 2100. The likely impact of these processes on marine plant life is currently a matter of hot debate. Rising temperatures will have impacts on metabolism and on processes such as macroalgal reproduction, which are likely to cause changes in geographical distributions of marine plants. In terms of impacts on photosynthesis, the changes are somewhat more contentious and variable. One predicted consequence of rising temperatures is for enhanced thermal stratification of the oceans, with a shallowing (shoaling) of the upper mixed layer. This has two consequences: For tropical and temperate latitudes, the major effect of increased stratification will be a reduced degree of mixing between the upper mixed layer and deeper waters, so that the supply of nutrients across the thermocline will be diminished. This will lead to enhanced nutrient limitation in the upper mixed layer and decreased photosynthesis and primary productivity. It is also likely to lead to shifts in phytoplankton communities away from larger species such as diatoms to smaller, picophytoplankton, such as *Synechococcus* and *Prochlorococcus*. Since smaller organisms sink less rapidly than large ones, this will in turn lead to a decreased export of carbon to the deep ocean and a weakening of the biological CO_2 pump in the oceans. At higher latitudes, where light is often the environmental factor limiting primary productivity, phytoplankton cells will find themselves retained in a shallower mixed layer where light is higher. Such waters may thus show higher rates of phytoplankton photosynthesis and primary productivity in the future.

The predicted changes in CO_2 in the atmosphere till year 2100 will translate to a 2.5- to 3-fold increase in dissolved CO_2 in seawater. This will cause the pH of the oceans to drop from pre-industrial values of ~8.2 to 7.7 (a phenomenon known as ocean acidification, see below). In turn, this decrease in pH will cause shifts in the equilibria between inorganic carbon forms (see Section 3.2) such that HCO_3^- concentrations will only rise by ~10% but CO_3^{2-} concentrations will decrease by ~50%. The total Ci available for photosynthesis ($HCO_3^- + CO_2$) will thus rise by <15%.

The above changes in the Ci system anticipated for the future might be expected to impact on species without CCMs more than those possessing CCMs as the latter show photosynthetic rates at, or close to, saturation by the present-day CO_2-equilibrated Ci concentrations. For phytoplankton though, the available data are not as clear cut as might be expected. Increased CO_2 above the present level has been shown to not significantly increase the specific growth rate of *Synechococcus* or *Prochlorococcus* whereas both CO_2 and elevated temperature increased the growth rate of *Trichodesmium*. Of six phytoplankton species assessed in the recent review of Doney et al. (2009), two showed increases in photosynthesis and four showed no change. Results of studies of coccolithophore and cyanobacterial photosynthesis under elevated CO_2 were equally split between no change and stimulation. Regarding macrophytes, it is sometimes suggested that many seagrasses are not CO_2 saturated for photosynthesis and, possibly, growth (see, however, in Section 7.4 about our doubts regarding this general statement). In cases where this is true, it is likely that the productivity of those species will increase in a future world. Since much of the increased productivity of seagrasses would be allocated to below-sediment biomass, carbon burial may increase with increasing carbon dioxide at saturating light, but not when light is limiting.

The increased acidity associated with higher CO_2 (ocean acidification, OA) is predicted to have severe consequences for calcifying organisms. Certainly OA has been shown to significantly decrease coral growth and calcification, and to affect negatively the growth of coralline red algae as well as calcareous green algae, but again the situation for calcifying microalgae such as coccolithophores is not clear cut, with decreases, increases and no change in calcification all being reported in the literature. Some of this variation is undoubtedly due to different experimental approaches, but some seems to reflect genuine differences between species and in some cases strains of the same species.

As has been shown above (e.g. in Section 11.3), the removal of CO_2 from seawater by photosynthesis and its release from respiration can, in dense macrophyte meadows, cause pH fluctuations (from pH ~7 to >10), especially under high irradiances. This has led to the suggestion that seagrass meadows, or kelp beds, could serve as refugia for, e.g. calcifying organisms, that may benefit from the higher average diel pH in such systems than in the open sea.

12.3 Scaling of photosynthesis towards community and ecosystem production

From this book it should be obvious that marine photosynthesis is quantitatively somewhat important on this planet. We have also discussed ways of measuring photosynthetic rates, and it would be good if those measurements could point more quantitatively towards productivity. So, yes, photosynthesis is the basis of plant growth, but no, it is not easy to use measurements of photosynthetic rates in order to quantitatively predict growth. There are several reasons for this: If photosynthesis is measured under controlled conditions in the laboratory, then of course those conditions are different from what the plants would encounter under natural growth conditions. But even if photosynthetic gas-exchange rates were to be measured *in situ*, several principal facts make the comparison with growth rates shaky. In rooted plants such as seagrasses, the whole-plant productivity includes the roots, but those are often not included in the measurements. In other plants (and also in seagrasses), there are also losses, e.g. by herbivory or shedding off senescing parts, which may be transported out of the plants' vicinity, and those are naturally not included in the measurements. Why then, do we spend a section on photosynthesis and productivity? First, our Publisher was unhappy that this book contained fewer pages than stated in the contract. Secondly, though, gas-exchange rates of whole marine plant communities **have** been measured and, together with *in situ* PAM fluorometry, those measurements show agreements both between one another and may relate to productivity estimates. A relevant example is given in Box 12.1 by Rui Santos who, together with co-workers (in a COST action entitled Seagrass Productivity: From Genes to Ecosystem Management, www.seagrassproductivity.com), compared photosynthesis and productivity measurements carried out on various levels of organisation in a seagrass meadow off Corsica.

Box 12.1   Photosynthesis and Productivity in a Seagrass Meadow

by **Rui Santos**, University of Algarve, PT (rosantos@ualg.pt)

Let us first agree on the definition of primary productivity as the rate at which light energy is converted into the chemical energy of organic compounds through photosynthesis. Why am I bothering you at this point of the book with basic definitions that were already stated in the introduction to Chapter 1? Just because when you increase the level at which you assess productivity, from the plant to the community or to the ecosystem level, the term used is, for some reason, no longer productivity. Do not blame me, but the rate of Ci incorporation into the autotrophs of a community is commonly called gross community production (GCP), so the P is not productivity but production (for the ecosystem level, which includes the water column above marine benthic plants, just exchange C by E). When the whole community respiration, CR (the rate at which CO_2 is released by the community), is taken into account, then we have the net community production (NCP). If NCP is positive (i.e. there is a net uptake of CO_2), this means that the community is behaving like a photosynthesising plant under favourable conditions, i.e. it is in an autotrophic metabolic state and thus it is a sink for CO_2. On the other hand, when NCP is 'negative', the community is heterotrophic and thus it is a source of CO_2. (For convenience, O_2 exchange can be used as a proxy for Ci balances.) To know the NCP and NEP is quite important in order to assess the role of marine plant communities as part of the ocean carbon budget.

So, in late 2011 a group of researchers (including two of this book's authors) gathered together in the wonderful STARESO marine station in Corsica (you must visit!) to try to understand how irradiance drives *in situ* photosynthesis of the seagrass *Posidonia oceanica* and how photosynthesis can possibly be scaled up to impact higher-order processes such as community and ecosystem production. A diagram is presented below (Figure 12.1) to guide you through what we investigated.

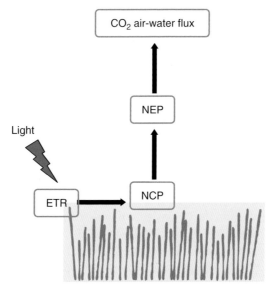

Figure 12.1　A schematic presentation of **production envisioned at various levels of organisation**, from the single seagrass plant leaf where electron transport rates (ETR) can be measured continuously by PAM fluorometry, through net community production (NCP) and net ecosystem production (NEP) measured by CO_2 or O_2 exchange in enclosed or open systems, respectively, to gas exchange rates between the water and overlying atmosphere. Drawing by Rui Santos.

Plant level: The diel photosynthetic production of *Posidonia oceanica* as driven by irradiance was assessed using the automated PAM fluorometers developed by John Runcie, which measure continuously and *in situ* the electron-transport rates (ETR) of multiple replicate samples (see Section 8.3 for PAM-fluorometric principles). These devices constitute a great improvement to the commonly used 'Diving-PAM', that only provides single point measurements, because diving time is a serious limiting factor to making many measurements. ETR *vs.* irradiance data were measured every 15 min for two days, providing a good model of the plant-level productivity response to light.

Community level: At the same time, we measured the diel variation of the whole community production of parts of the *Posidonia oceanica* meadow. NCP was derived from the O_2 evolution of a community section enclosed within benthic incubation chambers, slightly modified from those described in Santos et al. (see below). The water from the chambers was collected initially and at the end of incubations, which were restricted to 1.5–2.0 h to avoid depletion of CO_2 within the chambers (driven largely by increasing pH, see Section 3.2 for relationships between pH and the various Ci forms), which would consequently lead to an underestimation of NCP. The short duration of the incubations also allowed fitting a NCP *versus* irradiance function that was compared to the plant-level productivity model.

Ecosystem level: The production at the ecosystem level and the air-water CO_2 fluxes were also assessed. Both the O_2 and CO_2 concentrations in the water column were measured and logged at hourly intervals with Aanderaa optodes (at 4, 5, 7 and 9 m depth) and a Pro Oceanus analyzer (at 4.5 m), respectively, deployed in the water column over the *Posidonia oceanica* meadow that grew at 10 m depth. NEP was calculated by open-water mass balance of O_2 (after Odum, 1956). The air-water CO_2 fluxes were measured with the floating chamber method described by Frankignoulle (1988).

Plant vs. community: Within the irradiance levels to which the plants were subjected in the natural environment, we found a highly significant relationship between *Posidonia oceanica* photosynthesis (derived from the ETR *vs.* irradiance curves) and NCP, and that plant photosynthesis explained 82% of the variation of the whole community production. This was in spite of the many organisms that contribute to the community metabolism, both autotrophs that produce O_2 and heterotrophs that consume it, and the net metabolic contribution of which does not change significantly the contribution pattern of the seagrass' photosynthesis.

Community vs. ecosystem: The daily NEP, estimated in the water column over *Posidonia oceanica* community that grew at 10 m depth, scored between the daily NCP estimates made in meadows at 3 and 20 m depths. Furthermore, there was a highly significant relationship between the O_2 production by the community and the O_2 concentration in the water column. NCP explained about 84% of the variation of the NEP. The concentrations of CO_2 and O_2 in the water column over the *Posidonia oceanica* meadow were linearly, and inversely, related to each other as expected under the daily photosynthetic cycle. No relationships were found between the diel evolution of CO_2 in the water column and the air-water fluxes of CO_2, probably because these are affected by variables other than the differential CO_2 concentrations between air and water such as wind-driven turbulence of the liquid phase.

In this work, we established for the first time how a seagrass plant trait, photosynthesis, which is determined by the diel rhythm of irradiance, can scale up to impact higher-order processes such as community and ecosystem production. In fact, the strength of the relationships between the molecular electron-transport rate and the meadow community production, and between this and the whole ecosystem production, were surprisingly high. This highlights the key role of seagrasses themselves driving the coastal biogeochemical carbon fluxes at this, and probably other, seagrass-dominated systems. It also shows the potential for estimating production (or productivity, if we go back to its definition throughout this book). The relationships found here should thus be applied also to other marine plant communities so as to assess how general these observations are including other seasonal, spatial and species-specific contexts.

The following references were mentioned in this box: Frankignoulle M. 1988. Field measurements of air-sea CO_2 exchange. Limonology & Oceanography 33: 313–322; Odum HT. 1956. Primary production in flowing waters. Limnology & Oceanography 1: 102–117; Santos R. Silva J. Alexandre A. Navarro R. Barrón C. Duarte CM. 2004. Ecosystem metabolism and carbon fluxes of a tidally dominated coastal lagoon. Estuaries 27: 977–985.

Summary notes of Part III

- **Photosynthetic rates** are often measured as gas exchange of O_2 or CO_2. Such rates are **apparent** (or net) rates, and must be corrected for respiration if **true** (or gross) rates are sought. An alternative, and upcoming, way of measuring true photosynthetic rates (of electron transport through photosystem II, ETR) is by pulse-amplitude modulated (**PAM**) **fluorometry**. For phytoplankton, fast repetition rate fluorometry (**FRRF**) is also a useful alternative to PAM fluorometry.

- **Responses** of marine plants **to irradiance** can be expressed as photosynthesis *vs.* irradiance (*P–I*) **curves**, some of which (**rapid light curves, RLC**) can be generated within minutes using PAM fluorometry. In general, **low-light plants** show lower maximal rates of photosynthesis, but higher rates at low irradiances, than do **high-light plants**.

- Photosynthesis is sensitive to desiccation, but many **intertidal algae can 'shut down' their** **photosynthetic activity for extended periods of desiccation** and then revive it when resubmerged.

- Some marine plants can **modify the environment** in which they live so as to affect other plants either negatively (e.g. the **macroalga** *Ulva* can, by its photosynthetic traits, exclude other plants from rockpools) or positively (e.g. **seagrasses** can generate environments conducive to algal calcification and some **phytoplankton** can give off CO_2 during photosynthesis, which can possibly be used by others).

- The future holds **increasing CO_2 concentrations**, which may especially **favour those algae that lack an efficient CO_2-concentrating mechanism** (CCM), and possibly also some seagrasses. Increasing CO_2 concentrations will also **lower the seawater pH** ('ocean acidification'), which will be **detrimental for calcifying organisms** such as the calcareous algae.

The brown alga *Padina sp.* in the Mediterranean. Photo by Katrin Österlund.

References

Ahlgren G, 1987. Temperature functions in biology and their application to algal growth constants. Oikos 49: 177–190.

Badger M, 2003. The roles of carbonic anhydrases in photosynthetic CO_2 concentrating mechanisms. Photosynthesis Research 77: 83–94.

Badger MR, Andrews TJ, Whitney SM, Ludwig M, Yellowlees DC, Leggat W, Price GD, 1998. The diversity and coevolution of Rubisco, plastids, pyrenoids and chloroplast-based CO_2-concentrating mechanisms in algae. Canadian Journal of Botany 76: 1052–1071.

Beardall J, Giordano M, 2002. Ecological implications of microalgal and cyanobacterial CCMs and their regulation. Functional Plant Biology 29: 335–347.

Beardall J, Raven JA, 2004. The potential effects of global climate change on microalgal photosynthesis, growth and ecology. Phycologia 43: 26–40.

Beer S, Eshel A, Waisel Y, 1980. Carbon metabolism in seagrasses. III. Activities of carbon fixing enzymes in relation to internal salt concentrations. Journal of Experimental Marine Biology & Ecology 31: 1027–1033.

Beer S, Koch E, 1996. Photosynthesis of seagrasses vs. marine macroalgae in globally changing CO_2 environments. Marine Ecology Progress Series 141: 199–204.

Beer S, Vilenkin B, Weil A, Veste M, Susel L, Eshel A, 1998. Measuring photosynthetic rates in seagrasses by pulse amplitude modulated (PAM) fluorometry. Marine Ecology Progress Series 174: 293–300.

Beer S, Björk M, Gademann R, Ralph P, 2001. Measurements of photosynthetic rates in seagrasses. In: Short FT, Coles R (eds.) Global Seagrass Research Methods. Elsevier Publishing, The Netherlands, pp. 183–198.

Beer S, Björk M, Hellblom F, Axelsson L, 2002. Inorganic carbon utilisation in marine angiosperms (seagrasses). Functional Plant Biology 29: 349–354.

Björk M, Axelsson L, Beer S, 2004. Why is *Ulva intestinalis* the only macroalga inhabiting isolated rockpools along the Swedish Atlantic coast? Marine Ecology Progress Series 284: 109–116.

Björk M, Short F, Mcleod E, Beer S, 2008. Managing Seagrasses for Resilience to Climate Change. IUCN, Gland, Switzerland. 56pp. ISBN: 978-2-8317-1089.

Blankenship RE, 2002. Molecular Mechanisms of Photosynthesis. Blackwell Science, Williston VT, USA.

Borowitzka MA, Hallegraeff GF, 2007. Economic Importance of Algae. In: McCarthy PM, Orchard AE (eds.) Algae of Australia: Introduction. ABRS, Canberra, Australia.

Buapet P, Rasmusson LM, Gullström M, Björk M, 2013. Photorespiration and carbon limitation determine productivity in a temperate seagrass system. PLoS ONE 8(12): e83804. doi:10.1371/journal.pone.0083804.

Davison IR, Pearson GA, 1996. Stress tolerance in intertidal seaweeds. Journal of Phycology 32: 197–211.

Dawes CJ, 1998. Marine Botany. Wiley, Hoboken NJ, USA.

Doney SC, Fabry VJ, Feely RA, Kleypas JA. 2009. Ocean acidification: the other CO_2 problem. Annual Review of Marine Science 1: 169–192.

Drechsler Z, Sharkia R, Cabantchik ZI, Beer S, 1993. Bicarbonate uptake in the marine macroalga *Ulva* sp. is inhibited by classical probes of anion exchange by red blood cells. Planta 191: 34–40.

Drew EA, 1979. Physiological aspects of primary production in seagrasses. Aquatic Botany 7: 139–150.

Dring MJ, 1982. The Biology of Marine Plants. Edward Arnold Ltd., London.

Falkowski PG, Raven JA, 2007. Aquatic Photosynthesis. Princeton University Press, USA.

Gantt E, Cunningham FX, 2001. Algal Pigments. In Encyclopedia of Life Sciences. John Wiley, Oxford, UK. DOI: 10.1038/npg.els.0000323

Giordano M, Beardall J, Raven JA, 2005. CO_2 concentrating mechanisms in algae: Mechanisms, environmental modulation and evolution. Annual Review of Plant Biology 56: 99–131.

Glibert PM, and 56 others, 2008. Ocean urea fertilization for carbon credits poses high ecological risks. Marine Pollution Bulletin 56: 1049–1056.

Graham LE, Graham JM, Wilcox LW, 2009. Algae. Benjamin-Cummings Publishing Company, Zug, Switzerland.

den Hartog C, 1970. The Seagrasses of the World. North Holland Publ., Amsterdam.

Huot Y, Babin M, 2011. Overview of fluorescence protocols: Theory, Basic Concepts, and Practice. In: DJ Suggett, O Prasil, MA Borowitzka (eds.) Chlorophyll fluorescence in Aquatic Sciences: Methods and Applications. Developments in Applied Phycology 4: 31–74.

Hurd C, 2000. Water motion, marine macroalgal physiology and production. Journal of Phycology 36: 453–472.

Jassby AD, Platt T, 1976. Mathematical formulation of the relationship between photosynthesis and light for phytoplankton. Limnology & Oceanography 21: 540–547.

Johnston AM, 1991. The acquisition of inorganic carbon by marine macroalgae. Canadian Journal of Botany 69: 1123–1132.

Kirk JTO, 2011. Light and Photosynthesis in Aquatic Ecosystems. Cambridge University Press, Cambridge, UK.

Koch M, Bowes G, Ross C, Zhang X-H, 2013. Climate change and ocean acidification effects on seagrasses and marine macroalgae. Global Change Biology 19: 103–132.

Larkum AWD, Douglas SE, Raven JA (eds.), 2003. Photosynthesis in Algae. In: Govindjee (ed.), Advances in Photosynthesis and Respiration. Kluwer Academic Publishers, Dordrecht, The Netherlands.

Larsson C, Axelsson L, Ryberg H, Beer S, 1997. Photosynthetic carbon utilization by *Enteromorpha intestinalis* (Chlorophyta) from a Swedish rockpool. European Journal of Phycology 32: 49–54.

Legatt W, Marency E, Baillie B, Whitney SM, Ludwig M, Badger M, Yellowlees D, 2002. Dinoflagellate symbioses: Strategies and adaptations for the acquisition and fixation of inorganic carbon. Functional Plant Biology 29: 09–322.

Lobban CS, Harrison PJ, 1996. Seaweed Ecology and Physiology. Cambridge University Press, Cambridge, UK.

Losh JL, Young JN, Morel FMM, 2013. Rubisco is a small fraction of total protein in marine phytoplankton. New Phytologist 198: 52–58.

Mvungi EF, Lyimo TJ, Björk M, 2012. When *Zostera marina* is intermixed with *Ulva*, its photosynthesis is reduced by increased pH and lower light, but not by changes in light quality. Aquatic Botany, 102:44–49.

Papageorgiou GC, Govindjee (eds.), 2004. Chlorophyll Fluorescence: A Signature of Photosynthesis. Springer, Dordrecht, The Netherlands.

Platt T, Gallegos CL, Harrison WG, 1980. Photoinhibition of photosynthesis in natural assemblages of marine phytoplankton. Journal of Marine Research 38: 687–701.

Price GD, Badger MR, Woodger FJ, Long BJ. 2008. Advances in understanding the cyanobacterial CO_2-concentrating-mechanism (CCM): functional components, Ci transporters, diversity, genetic regulation and prospects for engineering into plants. Journal of Experimental Botany 59: 1441–1461.

Raven JA, Ball L, Beardall J, Giordano M, Maberly SC. 2005. Algae lacking CCMs. Canadian Journal of Botany 83: 879–890

Raven JA, Beardall J, 2003. CO_2 acquisition mechanisms in algae: Carbon dioxide diffusion and carbon dioxide concentrating mechanisms. In: Larkum A, Raven JA, Douglas S (eds.), Advances in Photosynthesis (Govindjee, Series ed.), Kluwer Academic Publishers, Dordrecht, The Netherlands.

Raven JA, Geider RJ, 1988. Temperature and algal growth. New Phytologist 110: 441–461,

Raven JA, Johnston AM, 1991. Mechanisms of inorganic-carbon acquisition in marine phytoplankton and their implications for the use of other resources. Limnology and Oceanography 36: 1701–1714.

Raven JA, Kübler JE, Beardall J, 2000. Put out the light, and then put out the light. Journal of the Marine Biological Association, UK, 80: 1–25.

Raven, JA, 2013. Rubisco: Still the most abundant protein in the world? New Phytologist 198: 1–3.

Raven PH, Evert RF, Eichhorn SE, 1992. Biology of Plants. Worth Publishers, New York.

Reinhold L, Kosloff R, Kaplan A, 1991. A model for inorganic carbon fluxes and photosynthesis in cyanobacterial carboxysomes. Canadian Journal of Botany 69: 984–988.

Richardson K, Beardall J, Raven JA, 1983. Adaptation of unicellular algae to irradiance: an analysis of strategies. New Phytologist. 93: 157–191.

Saffo MB, 1987. New light on seaweeds. BioScience 37: 654–664.

Saroussi S, Beer S, 2007. Alpha and quantum yield of aquatic plants as derived from PAM fluorometry. Aquatic Botany 86: 89–92.

Schreiber U, Bilger W, 1993. Progress in chlorophyll fluorescence research: Major developments during the past year in retrospect. In: Dietmar BH, Heidelberg UL, Darmstadt KE, Kederett JW (eds.), Progress in Botany. Springer-Verlag, Berlin.

Schwarz A-M, Björk M, Buluda T, Mtolera M, Beer S, 2000. Photosynthetic utilisation of carbon and light by two tropical seagrass species as measured *in situ*. Marine Biology 137: 755–761.

Semesi IS, Beer S, Björk M. 2009. Seagrass photosynthesis controls rates of calcification and photosynthesis of calcareous macroalgae in a tropical seagrass meadow. Marine Ecology Progress Series 382: 41–47.

Sharon Y, Silva J, Santos R, Runcie J, Chernihovsky M, Beer S. 2009. Photosynthetic responses of *Halophila stipulacea* to a light gradient: II – Plastic acclimations following transplantations. Aquatic Biology 7: 153–157.

Sharon, Y., Levitan, O., Spungin, D., Berman-Frank, I. and Beer, S. 2011. Photoacclimations of the seagrass *Halophila stipulacea* to the dim irradiance at its 48-m depth limit. Limnology & Oceanography 56: 357–362.

Steeman Nielsen E, 1975. Marine Photosynthesis. Elsevier, Amsterdam.

Taiz L, Zeiger E, 2010. Plant Physiology. Sinauer, Sunderland MA, USA.

Tcherkez GGB, Farquhar GD, Andrews TJ, 2006. Despite slow catalysis and confused substrate specificity, all ribulose bisphosphate carboxylases may be nearly perfectly optimized. Proceedings of the National Academy of Science 103: 7246–7251.

Tchernov D, Silverman J, Luz B, Reinhold L, Kaplan A, 2003. Massive light-dependent cycling of inorganic carbon between photosynthetic microorganisms and their surroundings. Photosynthesis Research 77: 95–103.

Uitz J, Claustre HG, Gentili B, Stramski D, 2010. Phytoplankton class-specific primary production in the world's oceans: Seasonal and interannual variability from satellite observations. Global Biogeochemical Cycles 24: GB3016, doi:10.1029/2009GB003680.

Wiencke C, Bischof K, (eds.), 2012. Seaweed Biology, Novel Insights into Ecophysiology, Ecology and Utilization (Ecological Studies, Vol. 219). Springer-Verlag, Berlin.

Williams P LeB, Laurens LML, 2010. Microalgae as biodiesel & biomass feedstocks: Review & analysis of the biochemistry, energetics & economics. Energy & Environmental Science 3: 554–590.

Winters G, Loya Y, Rottgers R, Beer S, 2003. Photoinhibition in shallow water colonies of the coral *Stylophora pistillata* as measured in situ. Limnology and Oceanography. 48: 1388–1393.

Winters G, Loya Y, Beer S, 2006. In situ measured seasonal fluctuations in Fv /Fm of two common Red Sea corals. Coral Reefs 25: 593–598.

Woodward FI, 2007. Global primary production. Current Biology 17: 269–273.

Yoon HS, Hackett JD, Ciniglia C, Pinto G, Battacharya D, 2004. A molecular timeline for the origin of photosynthetic eukaryotes. Molecular Biology and Evolution 21: 809–818.

Index

Photosynthesis in the Marine Environment, First Edition. Sven Beer, Mats Björk and John Beardall.
© 2014 John Wiley & Sons, Ltd. Published 2014 by John Wiley & Sons, Ltd.
Companion Website: www.wiley.com/go/beer/photosynthesis